AF823787

Green Energy and Technology

Climate change, environmental impact and the limited natural resources urge scientific research and novel technical solutions. The monograph series Green Energy and Technology serves as a publishing platform for scientific and technological approaches to "green"—i.e. environmentally friendly and sustainable—technologies. While a focus lies on energy and power supply, it also covers "green" solutions in industrial engineering and engineering design. Green Energy and Technology addresses researchers, advanced students, technical consultants as well as decision makers in industries and politics. Hence, the level of presentation spans from instructional to highly technical.

Indexed in Scopus.

Indexed in Ei Compendex.

Osamudiamen Otasowie · Clinton Ohis Aigbavboa ·
Ayodeji Emmanuel Oke

Circular Economy Business Model for Construction Organisations

Springer

Osamudiamen Otasowie
Sustainable Human Settlement
and Construction Research Centre
(SHSCRC)
University of Johannesburg
Johannesburg, South Africa

Ayodeji Emmanuel Oke
Department of Quantity Surveying
Federal University of Technology
Akure, Nigeria

Department of Construction Management
and Quantity Surveying
University of Johannesburg
Johannesburg, South Africa

Clinton Ohis Aigbavboa
Department of Construction Management
and Quantity Surveying, Sustainable
Human Settlement and Construction
Research Centre (SHSCRC) and CIDB
Centre of Excellence
University of Johannesburg
Johannesburg, South Africa

ISSN 1865-3529 ISSN 1865-3537 (electronic)
Green Energy and Technology
ISBN 978-3-031-88321-7 ISBN 978-3-031-88322-4 (eBook)
https://doi.org/10.1007/978-3-031-88322-4

© The Editor(s) (if applicable) and The Author(s), under exclusive license to Springer Nature Switzerland AG 2025

This work is subject to copyright. All rights are solely and exclusively licensed by the Publisher, whether the whole or part of the material is concerned, specifically the rights of translation, reprinting, reuse of illustrations, recitation, broadcasting, reproduction on microfilms or in any other physical way, and transmission or information storage and retrieval, electronic adaptation, computer software, or by similar or dissimilar methodology now known or hereafter developed.
The use of general descriptive names, registered names, trademarks, service marks, etc. in this publication does not imply, even in the absence of a specific statement, that such names are exempt from the relevant protective laws and regulations and therefore free for general use.
The publisher, the authors and the editors are safe to assume that the advice and information in this book are believed to be true and accurate at the date of publication. Neither the publisher nor the authors or the editors give a warranty, expressed or implied, with respect to the material contained herein or for any errors or omissions that may have been made. The publisher remains neutral with regard to jurisdictional claims in published maps and institutional affiliations.

This Springer imprint is published by the registered company Springer Nature Switzerland AG
The registered company address is: Gewerbestrasse 11, 6330 Cham, Switzerland

If disposing of this product, please recycle the paper.

Preface

The construction industry is significant in every economy. However, it is undeniable that the environment bears the brunt of the industry's activities. Key environmental impacts of the industry include resource depletion and waste generation. This is because of the linear economy business model adopted by the industry over the years. Circular economy (CE) focuses on preserving and enhancing materials and products through its principles, while business model (BM) is the process through which organisations create and deliver value. Therefore, the circular economy business model (CEBM) involves organisations creating value while aligning with CE principles. This book serves as a bridge between the theoretical potential of the circular economy business model and its practical application in construction organisations. It offers a unique chance to integrate CE principles into construction organisations' operations while promoting the efficient use and reuse of resources and reducing the reliance on virgin materials. Researchers and practitioners alike in the construction industry face the challenge of understanding both the technical aspects of the circular economy and how it can be integrated into the business model of construction organisations. With an emphasis on the exciting opportunities that lie ahead, this book seeks to guide researchers and practitioners along the trajectory. Through a progressive approach, this book demystifies the concepts of the circular economy and circular economy business models, highlighting their significance for construction organisations that should spearhead the adoption of this new concept. Each chapter builds on the previous one, beginning with an overview of the circular economy and business model and progressing to its practical applications, ultimately offering a comprehensive framework for implementation. Key themes throughout the book include circular value propositions, takeback systems, material information management systems, and organisational culture to ensure a closed-loop system for materials and components. Furthermore, instead of focusing solely on theoretical aspects, the book incorporates real-world applications and expert insights through a Delphi study. This approach will provide readers with a practical roadmap for integrating circular economy principles into their business models to improve material efficiency, recycling and reuse, economic development, and environmental protection.

Johannesburg, South Africa Osamudiamen Otasowie
Johannesburg, South Africa Clinton Ohis Aigbavboa
Akure, Nigeria Ayodeji Emmanuel Oke

Contents

Part I
Background Information

Chapter 1
Introducing Circular Economy in the Construction Industry

Abstract The construction industry, often labelled as one of the largest consumers and exploiters of natural resources, relies significantly on the environment for its raw materials. Construction activities significantly contribute to environmental pollution and degradation, impacting ecosystems from project inception to demolition. The critical environmental impacts of construction include resource depletion and waste generation. This is due to the linear business model that still dominates the industry despite a shift from the approach to a sustainable approach by other industries, like manufacturing. The circular economy (CE) is regarded as an alternative to the linear economy, which aims to extend the life of construction resources and reverse the typical propensity to take, make, and dispose of. Through resource-efficient design and usage, CE seeks to increase the lifespan of items and resources and eliminate waste. However, adopting CE in the construction industry requires a framework and model that conceptualises all CE principles to provide the roadmap for successful implementation. Hence, this book focuses on developing a CE business model to guide CE adoption in construction organisations. However, this chapter introduces CE in the construction industry by highlighting the evolution of CE and providing details on the study's context and the research problem. In addition, a basic overview of the study's significance and objectives are discussed in this chapter.

1.1 Introduction

The circular economy (CE) operates through various principles, including the 3Rs (recycle, reuse, reduce), 6Rs (redesign, reduce, reuse, recover, remanufacture, recycle), and 9Rs (reduce, refuse, repair, reuse, remanufacture, refurbish, recover, recycle, repurpose) [13]. CE strategies benefit organisations by addressing sustainability from economic, environmental, and social perspectives. The European Environmental Agency [12] notes that CE focuses on reducing emissions, improving resource use, and mitigating environmental impacts through recycling and renewable energy. CE also promotes remanufacturing, refurbishing, reusing, and shifting

© The Author(s), under exclusive license to Springer Nature Switzerland AG 2025
O. Otasowie et al., *Circular Economy Business Model for Construction Organisations*, Green Energy and Technology, https://doi.org/10.1007/978-3-031-88322-4_1

business models [9], which leads to economic gains, job creation, and resilience against raw material shortages [36].

Hence, replacing the current linear economy with CE could save billions while reducing environmental harm, especially by curbing resource extraction, waste, and emissions [9, 20]. The linear take-make-dispose model has increased waste and resource depletion, contributing to climate change and biodiversity loss [18, 20, 41]. These issues threaten fundamental human rights and economic stability [34, 37], prompting the need for CE to minimise raw material use and waste. The adoption of CE has spread across industries, driven by environmental regulations and sustainability goals. The agri-food sector, for example, has focused on reducing food waste and improving sustainability [14, 15, 38], while the fashion industry has embraced circularity through recycling and using biodegradable materials, with brands like Gucci and Balenciaga recycling materials to reduce waste [28]. Similarly, CE is being applied inconsistently in construction, which is a significant contributor to global GDP but also to resource depletion and environmental harm [11, 39].

CE is increasingly recognised for its potential in construction, though its implementation remains inconsistent [8]. The sector is crucial to the economy but also one of the most significant contributors to resource depletion and waste. Studies show that construction activities generate over 35% of the EU's waste and are responsible for significant greenhouse gas emissions [39]. Shifting to a CE model in construction focuses on extending resource lifespans through recycling, remanufacturing, and efficient design [42]. However, challenges to CE adoption include a lack of awareness, inconsistent understanding, and an entrenched linear business model in construction [21]. Thus, transitioning to CE requires new business models integrating sustainability into value creation [16, 29]. The CE business model (CEBM) aims to help organisations shift from linear to circular approaches by incorporating reuse, recycling, and resource efficiency into their operations [22]. This model enables organisations to retain resources in circulation for longer, minimising the need for new materials and reducing waste [5, 33, 40]. However, a lack of a comprehensive CEBM in construction has hindered CE adoption, emphasising the need for industry-specific frameworks that guide organisations through the transition to a CE [29]. Hence, this study focuses on developing such a framework for construction organisations.

1.2 The Evolution of Circular Economy

In the book chapter "*The Circular Economy*" by Pearce and Turner [30], the authors were among the first to discuss the concept of CE comprehensively. They introduced a flow diagram illustrating the input–output dynamics of materials, energy, and utility, challenging the traditional open linear economy to consider the interplay between the economy and the environment. This work recognised three essential economic roles for the environment: resource generation, waste absorption, and direct utility. They stated, "Economies exist, and the natural environment exists. What we do not know is what needs to occur for them to co-exist in equilibrium" [30, p. 42]. Following this, several economists in the 1990s began to embrace this perspective, recognising the shortcomings of existing economic theories in addressing environmental issues. In his paper "*Allocation, Distribution, and Scale: Toward an Economics that is Efficient, Just and Sustainable*," Daly [7] criticised mainstream economic theories for overlooking critical environmental challenges. He argued for the need to treat scale independently from allocation and distribution, likening the balance of these factors to that of a boat,if any element is unbalanced, the system fails.

At the same time, Industrial Ecology (IE) gained traction as a concept focused on minimising the environmental impacts of industrial processes that heavily rely on natural resources. Allenby, a prominent author in IE, contributed several works, including "*Achieving Sustainable Development through Industrial Ecology*" [1], which addressed the link between IE and sustainable development, and "*The Greening of Industrial Ecosystems*" [2], which examined IE in the context of materials science. Recent contributions from environmental economists have expanded these discussions, emphasising the necessity for minimal inputs and the efficient circulation of materials [3]. As IE developed, regenerative design emerged as a critical field in the built environment. Lyle [25] emphasised the idea of "waste as a resource" and proposed various strategies for regenerative design. However, he acknowledged that the practicality of utilising waste depends on its material properties. For instance, some polymers, like polyethylene (PET), can be recycled, while others, such as polypropylene (PP), pose challenges to closed-loop recycling [10]. Lyle also highlighted that regenerative materials must be economically viable and supported by public policy to be effective [25].

Furthermore, biomimicry, a subset of regenerative design, involves imitating biological processes at multiple levels, from individual organisms to ecosystems. Thus, Benyus [4] connected businesses to natural ecosystems, offering "ten lessons" to inspire ecologically conscious practices. Both regenerative design and biomimicry share a systems-thinking approach, considering holistic interactions rather than isolated features [44]. Also. Gunter Pauli's concept of the "*blue economy*" [32] serves as a significant foundation for CE, putting together ideas from IE, regenerative design, and biomimicry. In his early work, "*Zero Emissions: The Ultimate Objective of Cleaner Manufacturing*," he proposed innovative approaches to achieve zero waste through industrial symbiosis [31]. This idea was further elaborated in "*The Blue Economy: 10 Years, 100 Inventions, 100 Million Employments*," which presented practical solutions and emphasised integrated thinking [6, 32].

In addition, the influential book "*Factor Four*" by Weizacker et al. [43], presented to "*The Club of Rome*," highlighted the importance of integrated thinking in achieving sustainable economic growth, demonstrating that economies could quadruple resource output while halving energy consumption. This required not only the scaling of recommended technologies but also a holistic view of design challenges. McDonough and Braungart [26] "*Cradle to Cradle*" also emphasise integrated thinking, advocating for a new industrial revolution that harmonises production with nature. Natural capitalism, as outlined by Lovins et al. [23], further incorporates principles from natural ecology. They provided examples of how environmentally sound business strategies yield significant economic benefits. The three pillars of natural capitalism include enhancing the productivity of natural resources, operating in closed-loop systems, and developing solution-based business models [17]. The benefits of adopting these strategies are often exponential and achievable through minor design modifications and a holistic perspective.

Nevertheless, the quest to define CE formally began in the twentieth century, with Kirchherr et al. [19] analysing 114 definitions to arrive at their own, characterising CE as:

> …an economic system that replaces the "end-of-life" concept with reducing, alternatively reusing, recycling and recovering materials in production/distribution and consumption processes. It operates at the micro level (products, companies, consumers), meso level (eco-industrial parks) and macro level (city, region, nation and beyond), with the aim to accomplish sustainable development, thus simultaneously creating environmental quality, economic prosperity and social equity, to the benefit of current and future generations. It is enabled by novel business models and responsible consumers (pp. 224–225).

Hence, today's understanding of CE distinguishes itself from earlier interpretations through three fundamental shifts: first, it embraces systems thinking rather than mere efficiency [27], second, it advances beyond eco-industrial parks to include reverse logistics and closed-loop supply chains; enabling collaboration regardless of geographical proximity [24], and third, it expands the discourse from the traditional 3Rs to a broader framework of 10Rs [35].

1.3 Objectives and Contribution of the Book

The growing global demand for sustainable development emphasises the need for a safe, healthy environment and secure access to energy, health, food, and prosperity. To address this, the developed CEBM presented in this book offers a significant advancement for construction organisations seeking to integrate CE principles. The CEBM promotes resource efficiency and waste reduction, reducing dependence on virgin materials and optimising resource consumption, ultimately leading to cost savings. Hence, by embracing circular value propositions, such as manufacturing durable construction components, refurbishing existing items, and recycling materials, organisations can diversify revenue streams and contribute to several UN Sustainable Development Goals (SDGs), including SDG 1 (No Poverty), SDG 8 (Decent Work and Economic Growth), SDG 9 (Industry, Innovation, and Infrastructure), SDG 12 (Responsible Consumption and Production), and SDG 13 (Climate Action).

The construction industry's significant energy and resource consumption adversely affects environmental, social, and economic sustainability. A transition to

CE practices is essential, and this requires a comprehensive CEBM to guide organisations in incorporating these principles into their value-creation processes. This book aims to develop such a framework, helping construction organisations retain resources at their maximum value, minimising the need for virgin materials, and preventing materials from becoming waste when they are no longer useful. Thus, the objectives of this book include:

1. To identify factors that determine a CEBM for organisations in the construction industry.
2. To identify frameworks and theories for CEBM.
3. To determine the main and sub-attributes that influence CEBM for construction organisations.
4. To develop a CEBM for organisations in the construction industry.

1.4 Structure of the Book

1.4.1 Part I: Background Information

Chapter 1—This chapter discusses the introduction and highlights the evolution of circular economy. It provides details on the study's context and its primary research problem. In addition, a basic overview of the study's purpose and goals is included in this chapter.

1.4.2 Part II: Circular Economy Business Model in Organisations

Chapter 2 discusses the fundamental principles of circular economy and circular economy application in the construction industry. Chapter 3 discusses the theoretical perspectives on circular economy in organisations and policies, which includes the approaches and perspectives on circular economy in organisations and circular economy policies. Chapter 4 then focuses on the theoretical frameworks of circular economy business model, discussing the framework underpinning circular economy business model and developing a circular economy business model for construction organisations. Chapter 5 discusses the gaps in circular economy business model theoretical frameworks, which include material information management system and organisational culture, while Chap. 6 focuses on conceptualising the circular economy business model adoption framework, discussing core components of the framework, which are twelve in number, creating a systematic guide for implementation.

1.4.3 Part III: Case Study

Chapter 7 discusses the validation of the conceptual framework, which includes the selection of construction experts for a Delphi Study and the computation approach, while Chap. 8 presents the results of the Delphi study, which is based on the expert's opinions.

1.4.4 Part IV: Conclusion

Chapter 9—The study's conclusion is contained in this chapter.

1.5 Summary

The book's topic was introduced in this Chapter, which also included details on its organisation, background, and relevance. It also discussed the study's aim, objectives, and significance. Finally, it provided details on the format of the book.

References

1. Allenby BR (1992) Achieving sustainable development through industrial ecology. Int Environ Aff 4(1):56–68
2. Allenby BR, Richards DJ (1994) The greening of industrial ecosystems. National Academy Press
3. Andersen MS (2007) An introductory note on the environmental economics of the circular economy. Sustain Sci 2(1):133–140
4. Benyus JM (1997). Biomimicry: innovation inspired by nature: William Morrow
5. Bocken NM, De Pauw I, Bakker C, Van Der Grinten B (2016) Product design and business model strategies for a circular economy. J Ind Prod Eng 33(5):308–320
6. Bocken NM, Short SW, Rana P, Evans S (2014) A literature and practice review to develop sustainable business model archetypes. J Clean Prod 65:42–56
7. Daly HE (1992) Allocation, distribution, and scale: towards an economics that is efficient, just, and sustainable. Ecol Econ 6(3):185–193
8. Eberhardt LCM, Birkved M, Birgisdottir H (2022) Building design and construction strategies for a circular economy. Arch Eng Des Manage 18(2):93–113. https://doi.org/10.1080/17452007.2020.1781588
9. Ellen MacArthur Foundation (2015) Growth within: a circular economy vision for a competitive Europe. Ellen MacArthur Foundation 100:81–92
10. Eriksen MK, Christiansen JD, Daugaard AE, Astrup TF (2019) Closing the loop for PET, PE and PP waste from households: influence of material properties and product design for plastic recycling. Waste Manage 96:75–85
11. Erlandsson M, Levin P (2005) Environmental assessment of rebuilding and possible performance improvements effect on a national scale. Build Environ 40(11):1459–1471

12. European Environment Agency (2016) Diverting waste from landfill—effectiveness of waste management policies in the European Union. http://www.eea.europa.eu/publications/diverting-waste-from-landfill-effectiveness-ofwaste-management-policies-in-the-european-union/at_download/file (en)
13. Fonseca LM, Domingues JP, Pereira MT, Martins FF, Zimon D (2018) Assessment of circular economy within Portuguese organizations. Sustainability 10(7):2521
14. Food and Agriculture Organization of the United Nations (FAO) (2014) Save food: global initiative on food loss and waste reduction. Definitional framework of food loss. Food and Agriculture Organization of the United Nations. https://www.fao.org/3/i7657e/i7657e.pd
15. Hamam M, Chinnici G, Di Vita G, Pappalardo G, Pecorino B, Maesano G, D'Amico M (2021) Circular economy models in agro-food systems: a review. Sustainability 13(6):3453. https://doi.org/10.3390/su13063453
16. Hart J, Adams K, Giesekam J, Tingley DD, Pomponi F (2019) Barriers and drivers in a circular economy: the case of the built environment. Procedia Cirp 80:619–624
17. Hawken P, Lovins A, Lovins LH (1999) The next industrial revolution. In: Debating the earth: the environmental politics reader, pp 322–338
18. International Resource Panel (2019) Natural resources for the future we want. A report of the International Resource Panel. United Nations Environment Programme, Nairobi, Kenya. Retrieved from https://www.resourcepanel.org/reports/global-resourcesoutlook
19. Kirchherr J, Reike D, Hekkert M (2017) Conceptualizing the circular economy: an analysis of 114 definitions. Resour Conserv Recycl 127:221–232
20. Korhonen J, Honkasalo A, Seppälä J (2018) Circular economy: the concept and its limitations. Ecol Econ 143:37–46. https://doi.org/10.1016/j.ecolecon.2017.06.041
21. Kirchherr J, Piscicelli L, Bour R, Kostense-Smit E, Muller J, Huibrechtse-Truijens A, Hekkert M (2018) Barriers to the circular economy: evidence from the European Union (EU). Ecol Econ 150:264–272
22. Lewandowski M (2016) Designing the business models for circular economy—towards the conceptual framework. Sustainability 8(1):43
23. Lovins AB, Lovins LH, Hawken P (1999) A road map for natural capitalism. Harv Bus Rev 77(1):145–159
24. Lüdeke-Freund F, Gold S, Bocken NM (2019) A review and typology of circular economy business model patterns. J Ind Ecol 23(1):36–61
25. Lyle JT (1994) Strategies for regenerative design. John Wiley & Sons
26. McDonough W, Braungart M (2010) Cradle to cradle: remaking the way we make things. North Point Press
27. Mendoza JMF, Sharmina M, Gallego-Schmid A, Heyes G, Azapagic A (2017) Integrating backcasting and eco-design for the circular economy: The BECE framework. J Ind Ecol 21(3):526–544
28. Moorhouse D, Moorhouse D (2017) Sustainable design: circular economy in fashion and textiles. Des J 20(sup1):S1948–S1959. https://doi.org/10.1080/14606925.2017.1352713
29. Oluleye BI, Chan DW, Saka AB, Olawumi TO (2022) Circular economy research on building construction and demolition waste: a review of current trends and future research directions. J Clean Prod 357:131927
30. Pearce DW, Turner RK (1990) Economics of natural resources and the environment. Harvester Wheatsheaf
31. Pauli G (1997) Zero emissions: the ultimate goal of cleaner production. J Clean Prod 5(1–2):109–113
32. Pauli GA (2010) The blue economy: 10 years, 100 innovations, 100 million jobs. Paradigm Publications
33. Potting J, Hekkert MP, Worrell E, Hanemaaijer A (2017) Circular economy: measuring innovation in the product chain. Planbureau voor de Leefomgeving, (2544)
34. Raworth K (2017) Doughnut economics: seven ways to think like a 21st-century economist. Random House Business Books

35. Reike D, Vermeulen WJ, Witjes S (2018) The circular economy: new or refurbished as CE 3.0?—exploring controversies in the conceptualization of the circular economy through a focus on history and resource value retention options. Resour Conserv Recycl 135:246–264
36. Rizos V, Tuokko K, Behrens A (2017) The circular economy: a review of definitions, processes and impacts. CEPS Papers 12440, Centre for European Policy Studies. https://ideas.repec.org/p/eps/cepswp/12440.html
37. Rockström J, Steffen W, Noone K, Persson Å, Chapin FSI, Lambin E, Lenton T, Scheffer M, Folke C, Schellnhuber HJ, Nykvist B (2009) Planetary boundaries: Exploring the safe operating space for humanity. Ecol Soc 14(2):1–32
38. Secondi L, Principato L, Laureti T (2015) Household food waste behaviour in EU-27 countries: a multilevel analysis. Food Policy 56:25–40. https://doi.org/10.1016/j.foodpol.2015.07.007
39. Simpeh EK, Smallwood JJ (2018) Analysis of the benefits of green building in South Africa. J Constr Project Manage Innovation 8(2):1829–1851
40. Urbinati A, Chiaroni D, Chiesa V (2017) Towards a new taxonomy of circular economy business models. J Clean Prod 168:487–498
41. United Nations Environment Programme (2016) Global materialflows and resource productivity: Assessment report for the UNEP international resource panel. http://Unep.Org/Documents/Irp/16-00169_LW_GlobalMaterialFlowsUNEReport_FINAL_160701.Pdf
42. Upadhyay A, Akter S, Adams L, Kumar V, Varma N (2019) Investigating "circular business models" in the manufacturing and service sectors. J Manuf Technol Manag 30(3):590–606
43. Weizsacker EV, Lovins AB, Day R, Lovins LH (1997) Factor four: doubling wealth, halving resource use. Nature 389(6648):247
44. Wright D, Meadows DH (2008) Thinking in systems. Earthscan

Part II
Circular Economy Business Model in Organisations

Chapter 2
The Principles of Circular Economy

Abstract The Circular Economy (CE) has emerged as a transformative framework highlighting various approaches and goals to achieve sustainable development. It offers an alternative to the conventional linear model of "take-make-use-dispose," replacing it with a more sustainable cycle of "make-use-reuse-remake-recycle." This chapter explored the ten principles of CE, referred to as the "10 Rs," and introduced an additional "R" specific to the construction sector, emphasising the importance of design for disassembly as a critical topic in modern construction practices. The adoption of CE principles in managing construction and demolition waste aims to minimise waste generation while maximising the recovery, reuse, and recycling of materials from deconstructed buildings. These efforts contribute to reducing the environmental footprint of demolition activities, conserving finite natural resources, and cutting greenhouse gas emissions tied to waste disposal processes. Furthermore, integrating CE principles into the construction industry of developed countries has showcased a global commitment to sustainability. Although these nations face distinct challenges, they have made significant strides by implementing effective policies, advanced technologies, and collaborative efforts. In contrast, developing countries must address urgent needs by formulating comprehensive policies and strategies prioritising resource efficiency, waste reduction, and sustainable consumption practices. Overall, the insights gained reveal the potential of CE business models to drive more resilient, inclusive, and environmentally responsible societies, paving the way for sustainable progress across diverse contexts.

2.1 Introduction

The concept of the CE, while recently popularised, is rooted in older ideas and theories such as cradle-to-cradle, industrial ecology, and regenerative economics. CE has become an umbrella term encompassing various principles and practices. Its widespread use owes much to the Ellen MacArthur Foundation (EMF) work, which popularised it in the United Kingdom starting in 2013 [27]. However, the fundamental ideas behind CE trace back to the 1860s, with Simmonds [100] being

© The Author(s), under exclusive license to Springer Nature Switzerland AG 2025

O. Otasowie et al., *Circular Economy Business Model for Construction Organisations*, Green Energy and Technology, https://doi.org/10.1007/978-3-031-88322-4_2

among the first to advocate for strategies that mutually benefit the economy and the environment. Foundational literature addressing environmental issues began to emerge in the 1960s, further shaping the framework of CE.

The EMF offers a widely recognised definition of CE, which Korhonen et al. [61, 65] categorised into two main types. The first emphasises restorative and regenerative practices such as reuse, recycling, and recovery. The second group focuses on minimising virgin material use, promoting restoration, and achieving a sustainable economy through concepts like eco-efficiency. Schroeder et al. [97] define CE as an economic system that replaces the "end-of-life" model focusing on reducing, reusing, recycling, and recovering resources throughout production, distribution, and consumption. The goal is to create an industrial economy with restorative purposes, efficiently utilising materials, energy, labour, and information to regenerate natural and social capital [15, 27].

Fundamental to this model is the concept of "restorative" practices, which involve returning materials to their original economic applications or repairing their origins. However, the EMF's reports do not provide a singular, definitive description of CE. In contrast to the linear "take-make-dispose" economy, CE emphasises maintaining the value of materials through recycling and reducing the demand for waste disposal and raw material extraction, activities that contribute significantly to carbon emissions [88]. The term CE combines "circular," referring to the design of goods and services to slow, close, or narrow resource loops, and "economy," which involves industrial processes and financial outcomes [45]. Bocken et al. [16] highlight the need to shift organisations and stakeholders from linear thinking to circular strategies [73]. This approach also incorporates longevity in product design, emphasising efficient use of materials and energy while reducing dependency on virgin resources [52].

Many definitions of CE stress the importance of closed material loops, where materials are reused as components, products, or raw materials, and the processes enabling this, such as recycling and refurbishing [25, 27, 28]. Such loops are essential for minimising waste and reducing the extraction of resources from nature. For example, closed loops can involve soil restoration through organic waste recycling or the reuse of components across industries. Open-loop recycling, where materials enter a different production cycle, can also be beneficial if it avoids downcycling. Another aspect of CE is extending product lifespans, known as product life extension (PLE). Strategies like designing durable products for reuse help close resource loops and reduce waste. Conversely, creating robust products intended for short-term use does not align with CE principles [10, 34].

The EMF [27] outlines five key principles that underpin the concept of the CE: eliminating waste by design, driving resilience through diversity (balancing adaptability and efficiency), adopting renewable energy, applying systems thinking, and utilising cascading processes. The CE concept becomes a broad and dynamic framework when these principles are applied. Figure 2.1, adapted from the CE butterfly diagram, emphasises the critical role of systems thinking at its core. It illustrates how interconnected processes contribute to closing material loops, including multiple production stages, consumer activities, and recovery systems. This may involve five or six economic activities for simpler products, while more complex items require dozens or hundreds of steps. Successfully closing such loops requires a comprehensive understanding of the supply chain, ensuring no critical links are weak or missing. Hence, the relationships among various organisations within the supply chain are particularly significant, as they influence the functionality and integrity of the entire network. Systems thinking examines the complete structure, individual components, and their interconnections as an integrated whole [118]. This approach also highlights the importance of resilience, a system's capacity to recover from disturbances. Hence, a circular system reliant on a single connection is inherently fragile, as the failure of that link can disrupt the entire loop. Alternative pathways and contingency measures are essential to reducing such risks. While it is possible to close loops without these safeguards, over time, disruptions may still cause the system to fail, underlining the importance of resilience in sustainable systems design.

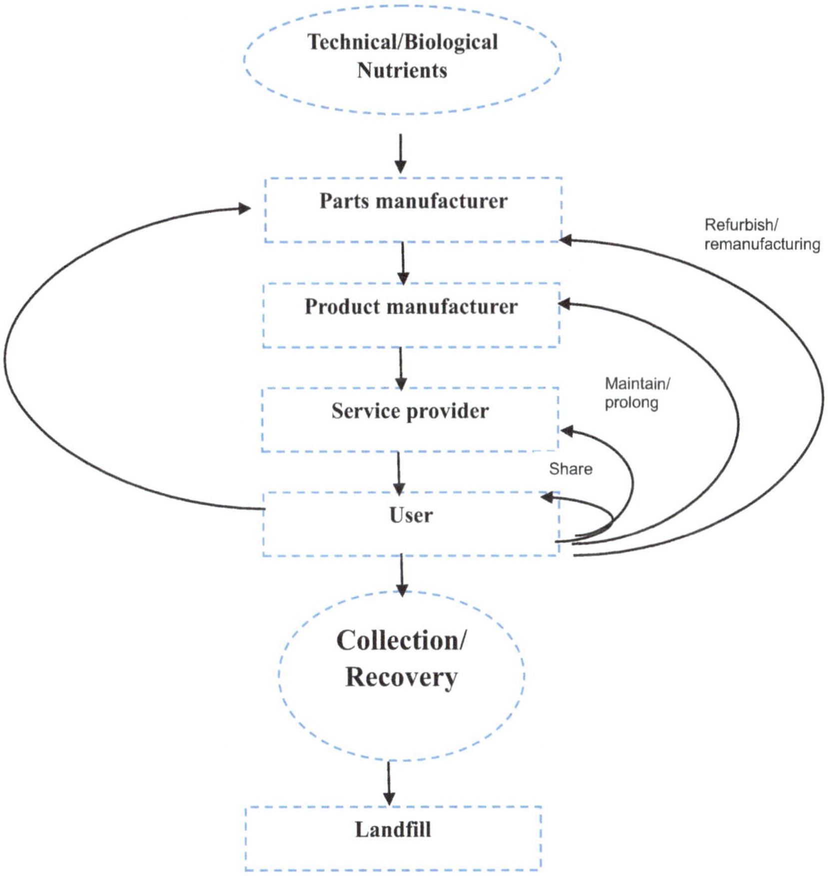

Fig. 2.1 Overview of the circular economy in a schematic form (adapted from EMF [27])

2.2 Key Principles of Circular Economy

According to Fonseca et al. [41], the guiding principles of CE have been a journey from the 3Rs (recycle, reduce, reuse) to the 6Rs (redesign, reduce, reuse, recover, recycle, remanufacture) and then to the 9Rs (reduce, refuse, repair, reuse, remanufacture, refurbish, recover, repurpose, recycle). In addition, Çimen [19] introduced 'replace' to the core principles of CE.

2.2.1 *Reduce*

The "reduce" approach focuses on designing new products, processes, or business models that substitute nonrenewable raw materials with recyclable, renewable, or biodegradable resources while manufacturing new products [107, 108, 123]. This method enables recycling organic waste, such as animal manure or decaying vegetables, into sustainable and eco-friendly alternatives, preserving nonrenewable natural resources. Goyal et al. [51] highlighted Haathi Chaap as a notable example of this approach. The company innovatively utilised elephant dung as a raw material to manufacture paper and related products. Haathi Chaap distinguished itself within the handmade paper industry by transforming production methods and resource usage. Its minimal setup costs for production equipment further underscored the efficiency of this model. In addition, the leftover water from washing elephant dung was repurposed as organic fertiliser in developed regions [51]. The company leveraged its access to abundant raw materials to scale operations and provide significant employment opportunities in smaller towns. According to Goyal et al. [51], the production process involves stages such as washing, boiling, sanitising, sorting, pulping, pressing, and drying to produce finished paper sheets. These products were then distributed through established retail networks.

The concept of "reduce" can be understood in three contexts: broadly, as a consumer- or producer-oriented term, or both [91]. Francis et al. [42] described it broadly as eliminating waste generation at its source rather than focusing on disposal after production. While this applies to all lifecycle stages, including usage, consumer responsibilities during this phase are often not explicitly outlined [47]. Reike et al. [91] suggested that consumers could extend product lifespans through careful usage, repairs, or collaborative repair activities, such as repair shops. Similarly, Den Hollander and Bakker [21] highlighted participation in the sharing economy through concurrent or sequential product use to maximise efficiency. On the production side, "reduce" typically emphasises less material usage per unit, with dematerialisation as a key step in design and development [56, 68, 94, 99, 116].

Adopting the "reduce" strategy in the construction sector offers significant potential to reduce raw material consumption, conserve natural resources, and reduce carbon emissions [27, 28]. Among the "R" strategies, "reduce" is often considered the most environmentally sustainable approach to managing construction and demolition waste (CDW). Thus, it is prioritised in CDW management plans [54]. Kabirifar et al. [63] stated that the initial focus should be minimising waste generation. If waste cannot be avoided, the following steps involve reusing or recycling materials before disposal. Pickin et al. [87] outlined the benefits of reducing waste, including cost savings from reduced material purchases, revenue from collecting specific waste items, lower CO_2 emissions, and reduced expenses for transporting waste to landfills. Emphasising the reduce strategy within the CDW management cycle ensures that construction stakeholders maximise these benefits, especially given the global increase in CDW generation [31].

2.2.2 Reuse

The reuse paradigm focuses on extending a product's lifespan while repurposing discarded materials to fulfil essential societal needs and generate economic value. This approach drives a "waste" economy that prevents valuable resources and materials from ending in landfills while creating alternative income streams [32, 46, 117]. An illustrative example of this paradigm is Goonj. This company developed a unique business model to empower rural villagers, especially women, by offering them opportunities to earn income, enhance their skills, and contribute to their community's infrastructure. Goonj recognised that a significant portion of India's rural population lacked access to affordable clothing, and 68% of rural women were deprived of hygienic sanitary products [17]. By establishing a network that redirected surplus resources from urban households to underserved rural and disaster-stricken communities, Goonj enabled resource distribution while facilitating skill development, community organisation, and infrastructure projects in exchange for essentials such as clothing and household items [35].

In the construction sector, reusing construction and demolition waste (CDW) involves repurposing materials for their original function or a different application [54]. Most CDW materials can be reused after deconstruction, making reduction and reuse essential strategies for environmental conservation, resource preservation, and cost savings. Reusing construction waste also supports sustainable practices by reducing greenhouse gas emissions and extending the lifespan of materials [82, 84]. Materials from demolition sites can often be salvaged for future use, sold, or incorporated into ongoing projects. However, certain components like chemical solvents, adhesives, and latex paints pose hazardous risks and require cautious handling [82].

Reusing industrial by-products is another promising strategy. Xiao et al. [119] and Skaf et al. [102] highlighted that such by-products can be employed in road construction, replacing natural aggregates in hydraulically bound composites for road foundations without compromising quality or functionality. This not only reduces landfill disposal but also introduces sustainable practices. Similarly, Allwood et al. [8] noted that 40% of steel used in construction, such as structural sections and reinforcing bars, can be reused, provided designs accommodate part reuse, aligning with CE principles.

Studies have emphasised the significance of incorporating reuse strategies in construction waste management. Kamaruddin and Marwan [64] advocated for integrating reuse into the design phase of projects to encourage resource-efficient construction practices. In a Dutch case study, Nguyen et al. [74] demonstrated that government policies can support CDW reuse through incentives and the development of recycled material markets. Kabirifar et al. [63] stressed the role of construction companies in promoting reuse by forming partnerships with recyclers and sorting materials for recovery. Umar et al. [109] highlighted the potential for reuse in structural and non-structural applications, suggesting that governments could enhance these efforts by offering incentives, developing standards for recycled materials, and mandating the use of such materials in public projects. Together, these efforts underscore the importance of reuse in conserving resources, minimising waste, and mitigating the environmental impacts of construction activities.

2.2.3 Recycle

The recycling model focuses on actively reducing waste by transforming it into clean resources, thereby balancing the creation and consumption of materials [107, 123]. This approach emphasises collecting and recycling solid and electronic waste to minimise material leakage and enhance economic value to create new products [46, 107]. Recycling is typically classified into two types: closed-loop and open-loop. Closed-loop recycling reuses waste to manufacture new products while maintaining the original material's properties, whereas open-loop recycling uses recovered materials to create lower-value products [107, 108]. Within the framework of a CE, recycling involves establishing a comprehensive process to collect, sort, and classify electronic waste for extracting usable metals. Global quality standards, skilled personnel, and advanced technology support this process.

For CDW, recycling involves disassembling materials from construction sites to create new materials. However, challenges such as inadequate recycling management, immature markets for recycled products, and outdated recycling technologies can hinder progress [54]. CDW recycling can be conducted onsite or offsite, depending on project capabilities and facilities. Materials like concrete, asphalt, wood, metal, plasterboard, corrugated cardboard, and roofing materials can be recovered from demolition sites [63]. Recycling construction materials also offers environmental benefits, such as significantly reducing CO_2 emissions [82]. Additionally, CDW recycling minimises the need for virgin material extraction, minimises pollutants and greenhouse gas emissions, conserves landfill space, and decreases energy consumption and environmental degradation [66, 87]. Furthermore, recycling generates economic activity and employment opportunities in related industries, using recycled materials for roads, foundations, sports fields, noise barriers, and landscaping [36].

Research indicates that recycled aggregates in construction have a lower environmental impact by reducing waste and the demand for raw materials. They also improve concrete properties and reduce costs, making them a cost-effective choice [11, 53]. Recycled gypsum, derived from discarded drywall, can be used to manufacture new drywall or other construction materials, reducing the need for virgin gypsum and decreasing waste. This practice can significantly cut greenhouse gas emissions associated with gypsum production [101, 114]. Another approach to CDW recycling is deconstruction, which involves carefully dismantling buildings to salvage reusable materials. Deconstruction can divert up to 90% of building materials from landfills, substantially reducing waste and conserving resources [58]. Globally, CDW accounts for about 30% of waste production [33], yet it plays a crucial role in supporting social and economic development while delivering economic and environmental benefits [14, 49]. Recycling is an indispensable element of CDW management. CDW recycling reduces waste, conserves resources, and minimises environmental harm through recycled aggregates, gypsum recovery, and deconstruction.

2.2.4 Refuse

"Refuse" encompasses both consumer and production waste. In the consumer context, it emphasises the decision to reduce consumption, aiming to avoid waste generation by intentionally buying or using less [8]. This reflects a shift in consumer values towards minimalism and a critical outlook on materialism. For manufacturers, "refuse" pertains to designing product life cycles that exclude harmful components and adopt waste-minimising production methods [13]. As a fundamental principle of the CE, "refuse" focuses on preventing waste by avoiding unsustainable or non-recyclable materials and practices. This can involve creating products for reuse or opting for eco-friendly and renewable materials. "Refuse" may also extend to forgoing certain products or achieving the same function through alternative, sustainable solutions.

For instance, avoiding non-recyclable materials in the construction industry represents a significant stride toward sustainability and mitigating the environmental impact of construction CDW. Utilising non-recyclable materials depletes natural resources and escalates waste production, whereas opting for recyclable alternatives conserves resources and minimises waste. In addition, incorporating recyclable materials in construction can save costs by reducing energy consumption and production waste [29]. By adopting such practices, construction firms can contribute to environmental sustainability and reduce material costs.

2.2.5 Repair

The concept of "repair" can sometimes be misunderstood despite its clear purpose: extending the lifespan of products [60]. It is described as "bringing back to working order" [38, p. 194], "making it as good as new" [103, p. 540], "restoring its original function after minor defects" [104, p. 222], or "replacing broken parts." Repair plays a vital role in CE by prolonging product usability, reducing the need for new production, minimising waste, and conserving resources. By keeping products in use longer, repair significantly lowers waste generation. For example, EMF [28] opines that extending a product's life by just nine months can cut its carbon, water, and waste footprint by 20–30%. This highlights the environmental benefits of repair in promoting sustainability.

Although repair might seem straightforward, its interpretation can vary depending on the context. In some cases, it overlaps with what others term "refurbishment," such as in discussions of military closed-loop systems [115]. A key attribute of repair is that it can be performed by different entities, such as the product owner, local service providers, or professional repair firms, and does not necessitate a transfer of ownership. Commercial repair workshops have recently gained traction [26], and companies increasingly repair returned goods at specialised facilities.

In CDW management, repair involves salvaging and reusing materials damaged during demolition. This reduces waste generation and the demand for new materials, supporting sustainability and lowering environmental impacts. Repair also offers economic benefits; for example, the Waste and Resources Action Plan [120] found that repairing and reusing materials can save up to 20% compared to purchasing new ones. This is particularly significant given rising raw material costs and the push for sustainable construction practices. Promoting repair in CDW management can be achieved through strategies such as designing buildings for durability, implementing salvage and reuse programs at demolition sites, and supporting repair and refurbishment industries [20].

2.2.6 Remanufacture

"Remanufacture refers to the process of disassembling, inspecting, cleaning, and, if required, replacing, or repairing an item's whole structure as part of an industrial process" [68: 44]. This process extends the life span of items that might otherwise be discarded, offering environmental, societal, and economic benefits [57]. Hazen et al. [55] highlight that remanufactured products require approximately 85% less energy compared to manufacturing equivalent new items. In addition to environmental advantages, remanufacturing positively impacts society by creating opportunities for both skilled and unskilled labour and supports the economy by offering cost-effective alternatives to new goods [69]. Typically, remanufactured products are sold at around 60% of the price of new items, making them an attractive option for consumers [89].

This practice can significantly reduce waste, potentially cutting construction sector waste by up to 70% [105] while minimising the need for virgin materials, conserving natural resources, and lowering environmental impacts. For example, substituting reclaimed bricks for new ones can cut energy consumption by as much as 50% [121]. In CDW management, remanufacturing offers several practical applications, such as reclaiming and reusing bricks after cleaning and sorting [121], salvaging steel from demolished structures for use in new projects [105], processing demolished wood for reuse in new construction [86], and crushing concrete to create aggregates for new concrete production [23]. These strategies contribute to sustainable construction practices by reducing waste, conserving resources, and promoting the circular use of materials.

2.2.7 *Refurbish*

The term "refurbish" is most appropriate for cases where the general structure of a complex, multi-component product is retained. Still, several parts are repaired or replaced, resulting in an overall upgrade [38]. Hence, "refurbishment" can also apply to the renovation of facilities. In CE, examples of products that are often refurbished include mining shovels, locomotives, aeroplanes, and machines. These refurbished items are typically brought up to modern standards using advanced components [70, 104]. Thus, refurbishment can also refer to the process of upgrading an existing building rather than demolishing it and rebuilding it from the ground up. Regarding CDW management, refurbishment is crucial in minimising waste and reducing environmental impact. Tanskanen et al. [106] explored the potential of refurbishment in Finland as a strategy to reduce CDW, finding that it could lower waste generation by as much as 60% compared to demolition and new construction. Also, the study highlighted that refurbishment could result in lower energy consumption and reduced greenhouse gas emissions. Ferreira et al. [37] found similar benefits in China, including a 50% reduction in waste generation and considerable cost savings compared to new construction. Refurbishment also has cultural and social advantages, as it helps conserve historically significant buildings and provides job opportunities in the construction sector [12].

2.2.8 *Recover*

The word 'recover' has a variety of uses, with three forms predominating, such as some of the other Rs. It first refers to "gathering old items at the end of their useful lives, followed by disassembly, sorting, and cleaning for usage" [122: 98]. This definition is commonly seen in reverse logistics literature. It may also involve extracting components or materials from composites that are no longer in use [104]. The primary aim of product recovery is to close the loop at every stage of the product's life cycle. Additionally, recovery seeks to maximise the economic and ecological value of discarded goods, parts, and materials, thus reducing waste sent to landfills.

Material recovery plays a crucial role in sustainable waste management in CDW management. It reduces the volume of waste sent to landfills and supports CE by enabling the recycling and reuse of materials. CDW materials that can be recovered include concrete, bricks, metals, wood, plastics, and other components, all of which can be sorted, cleaned, and repurposed for new products or construction applications. Recovering these materials offers both environmental and economic advantages, such as lessening the need for virgin materials, conserving energy, cutting greenhouse gas emissions, and driving job creation in the recycling and construction sectors. For example, recycling concrete can lower CO_2 emissions by up to 30% compared to producing new concrete [33], while recycling wood waste can conserve up to 50% of the energy used in making new wood products [110]. Thus, recovering materials from CDW transforms waste into valuable resources, helping to reduce overall waste and generate economic benefits.

2.2.9 Repurpose

The concept of 'repurpose' is often overlooked, as many studies do not specifically include it within the set of Rs. Li [67] suggests that when other authors mention 'rethink,' they may refer to the same idea. In fields like industrial design and art, repurposing is a common practice. This involves reusing discarded items or components and repurposing them for different functions, which gives the materials a new life. Such processes can apply to both low-value and high-value products. Stahel [104] highlights examples of unemployed individuals who started their businesses by transforming damaged microchips into jewellery, glass bottles into mugs, textile waste into quilts, or plastic sheeting into purses.

Repurposing plays a significant role in waste management, particularly concerning CDW. One prominent method of repurposing CDW is recycling. This involves converting waste into new products or raw materials suitable for constructing new buildings. For instance, crushed concrete can serve as aggregate in new concrete mixes, while metals can be melted down and reformed into new steel products. Another approach involves using reclaimed materials and building components salvaged from demolished structures. These materials can be repurposed for new construction, granting them a second life. Reclaimed materials might include bricks, timber, and architectural features like doors and windows.

2.2.10 Replace

The 'replace' concept in CDW management involves substituting existing materials with more sustainable, cost-effective, and environmentally friendly alternatives

[33]. It is defined as "the substitution of traditional materials and techniques by more sustainable, innovative, and efficient ones to enhance resource efficiency and reduce environmental impact" [33: 2]. This approach aims to lessen the environmental footprint of demolition by reducing waste generation and increasing material reuse. An example of this is the use of recycled aggregates in place of conventional construction materials like concrete and asphalt. Recycled aggregates are derived from processed CDW and can be applied in various construction activities, including road construction, landscaping, and as foundational material for new buildings [24]. Another example of 'replace' is using eco-friendly insulation materials instead of traditional options like fibreglass and foam. Materials such as cellulose, wool, and recycled denim offer benefits over conventional insulation, including better thermal efficiency, a smaller environmental footprint, and improved indoor air quality [62].

Çimen [19] notes that the 'replace' approach primarily targets the materials used in production. It focuses on introducing sustainable substitutes, even though these alternatives may increase economic value or reduce raw material consumption as secondary benefits. The 'replace' strategy might require using more raw materials than previously, but it still significantly lowers environmental impact. By applying this principle alongside other "R" strategies, additional benefits and outcomes can be further distinguished [19].

2.2.11 Reassembly

In this study, the concept of reassembly is integrated into the core principles of CE for the construction industry, although it can also be applied to the manufacturing sector. The reassembly process involves detailed planning, selective dismantling, and careful materials handling to maximise their value [3]. While reassembly shares similarities with remanufacturing, it is more appropriate for the construction industry, especially considering the growing emphasis on design for disassembly as a key aspect of the design process [3]. Reassembly is a method of managing CDW that focuses on dismantling structures and recovering materials for reuse, recycling, or resale. According to Kibert et al. [59], this process can recover up to 95% of building materials, which helps minimise the environmental impact of demolition and reduces costs compared to conventional methods that rely on disposal. The process demands careful planning, selective disassembly, and skilful material handling to ensure the materials retain their integrity and are reused effectively. Skilled labour and specialised equipment are necessary to prevent damage during dismantling. The U.S. Environmental Protection Agency [110] notes that reassembly requires a structured approach, which includes site assessments, material inventories, salvage planning, and proper storage.

Reassembly offers several advantages for sustainable development. It helps create job opportunities and supports local economic growth by providing recycled materials for construction projects [29]. Additionally, it reduces the demand for virgin materials, lowering environmental impacts such as energy consumption, carbon emissions, and resource depletion. Furthermore, it can cut waste disposal costs and contribute to environmental justice by easing the waste burden on low-income communities [29]. However, for reassembly to be successful, it is essential to adopt the right design strategies, such as design for disassembly. Also known as reverse construction, design for disassembly simplifies the deconstruction steps through thoughtful design and planning. Deconstruction involves dismantling a structure to recover materials for reuse or recycling [93]. This process is critical when a building reaches the end of its lifespan, as it promotes the reuse of components and reduces waste. Implementing design for disassembly involves rethinking traditional planning, design, and construction methods. It includes creating easily disassembled connections, keeping accurate records of components and materials, and designing parts with simple architecture and standardised dimensions. Hence, modular construction can facilitate disassembly and reassembly within the construction industry [93].

Modular construction involves the off-site fabrication of structural units, which are transported and assembled on-site to form larger, permanent structures [96]. In modular buildings, deconstruction may involve separating the modules once the building has reached the end of its useful life, allowing the modules to be reused at a different location. In contrast, recycling involves disassembling individual components of the modules, such as wall cladding, non-structural panels, flooring, and finishes, to extract clean, reusable materials [76]. This approach helps drive circularity and reduce CDW, contributing to a more sustainable construction industry through disassembly and reassembly.

2.3 Circular Economy Application in the Construction Industry

The unchecked growth of solid waste, driven by population increases and the consumption patterns of profit-driven economies, underscores the relevance of the current linear resource model, which involves the extraction, use, and disposal of resources through landfills [28]. However, this model can lead to significant environmental, economic, and social challenges if ecosystems and waste management structures fail [61, 65]. In response, global researchers and practitioners are exploring ways to minimise waste generation and recover value from waste materials, guided by the need to protect the environment. This shift is leading to a focus on sustainable production-consumption models, such as the CE, which aims to recirculate resources through various supply chains to extract as much value as possible, all while ensuring proper disposal to reduce pollution [28].

The construction sector is a major global consumer of raw materials [30]. In the United States alone, 569 million tons of CDW were generated in 2017 [112], and global CDW is projected to reach 2.2 billion tons by 2025 [30]. While the potential for recycling and reusing construction materials is vast, only about 40% of C&D waste is currently recycled, reused, or sent to waste-to-energy facilities, with the remaining 60% disposed of in landfills [111]. However, countries like the Netherlands are leading the transition to a fully circular economy by 2050, with the construction sector playing a crucial role in reducing CDW and implementing circular design principles. Initiatives such as modular construction, building component reuse, and sustainable materials are helping drive this transition [92]. Notable projects, like the "Circl Pavilion" in Amsterdam, showcase modular wooden panels and demountable steel frames as part of circular construction [85]. In addition, innovative models like "Material Passports" enhance transparency and lifecycle tracking, aiding the CE shift [113].

Italy has also made significant progress in adopting CE principles in its construction sector, driven by EU waste management directives and national sustainability policies. The country has focused on preserving cultural heritage while implementing CE practices, such as using non-toxic materials and material passports, which help minimise environmental impact [6, 48]. Similarly, the UK is advancing CE integration in construction, utilising Building Information Modeling (BIM) to optimise resources and enhance efficiency. The country's construction projects use recycled materials, and designs focused on material recovery. In leveraging its cement industry, Portugal has developed low-carbon, recycled-content materials to reduce environmental impacts [18].

In contrast, many developing nations face significant challenges with solid waste management (SWM), often relying on landfills or open dumps, which pose environmental and social risks [7, 39, 71, 90]. These countries often lack adequate resources, infrastructure, and maintenance facilities, making sustainable waste management a relatively new and challenging concept [83]. In Africa, the concept of CE remains underdeveloped in research, practice, and policy [22], with most environmental and economic initiatives still aligned with a linear economic model. However, interest in CE is growing across Africa to promote sustainable development [72]. Recycling is a widely accepted CE practice in Egypt, potentially linked to the country's extensive renewable energy projects, including wind and solar energy [44]. Ghana, facing pressure to adopt sustainable building techniques, has gradually embraced eco-friendly practices, including CE principles [43]. Many construction companies in Ghana follow the CE's 6R framework (repair, reduce, reuse, recover, recycle, remanufacture), though practitioners' understanding of these concepts remains limited [5, 9]. Ghana's construction industry still faces challenges in implementing environmentally sustainable practices due to the government's and stakeholders' incoherent environmental plans [75].

Tanzania is making strides in recycling CDW, such as concrete, into aggregates for new concrete production [95]. While concrete recycling is typical, there is potential for increased awareness and use of CE in the construction sector if CDW proves successful in new building applications [50]. In Nigeria, the growing waste from building materials has led to more significant pressure on the construction sector to reduce waste and close material loops [77]. Though waste management strategies such as prefabricated components, waste sorting, and reuse are in place, much of the material waste is still disposed in landfills [4, 78, 98]. However, there are examples of successful waste reuse practices, such as recycling wood, steel, and tile offcuts on-site or repurposing materials like concrete waste for roadwork and other uses [1,

2]. Lastly, South Africa faces a similar challenge, with more than a million tons of building and demolition waste sent to landfills annually. However, recycling remains a key waste management strategy, with practices such as using reclaimed asphalt for road pavement layers and repurposing timber and plastic for new products [40, 79–81]. Thus, the benefits of recycling under the CE model include reducing the need for new resources, saving energy, and minimising landfill waste, contributing to a more sustainable construction sector.

2.4 Summary

This chapter explored the ten (10) Rs of Circular Economy (CE) principles, highlighting the inclusion of a new "R" that is particularly relevant to the construction sector. This is due to the increasing importance of designing for disassembly in the industry's design process. Additionally, the chapter examined the implementation of CE principles within the construction industries of both developed and developing nations. The fundamental takeaways emphasise the potential of CE business models to drive resilient, inclusive, and environmentally sustainable societies.

References

1. Abdullahi AL, Lee A (2018) Is the concept of waste universal? Handling building demolition by-products in the city of Kano, Nigeria. J Constr Project Manage Innov 8(1):1779–1788
2. Abdullahi AL, Otasowie K, Lee A, Awuzie BO, Aigbavboa C, Oke A (2023) Conceptualising an ethno-mimetic model for effective buildings' end-of-life waste management: a Nigerian exemplar. Bus Strat Dev 6(3):322–332. https://doi.org/10.1002/bsd2.241
3. Abuzied H, Senbel H, Awad M, Abbas A (2020) A review of advances in design for disassembly with active disassembly applications. Eng Sci Technol Int J 23(3):618–624
4. Adewuyi TO, Odesola IA (2016) Material waste minimisation strategies among construction firms in south-south, Nigeria. Int J Sustain Constr Eng Technol 7(1):11–29
5. Adinyira E, Agyekum K, Amudjie J, Salgin B, Darko A (2022) Construction and the circular economy. In: Adinyira E, Agyekum K (eds) The construction industry: global trends, job burnout and safety issues. Nova Science Publishers, pp 75–104
6. Alhawamdeh M, Lee A, Saad A (2024) Designing for a circular economy in the architecture, engineering, and construction industry: insights from Italy. Buildings 14(7):1946
7. Al-Khatib IA, Monou M, Zahra ASFA, Shaheen HQ, Kassinos D (2010) Solid waste characterization, quantification and management practices in developing countries. A case study: Nablus district–Palestine. J Environ Manage 91(5):1131–1138
8. Allwood JM, Ashby MF, Gutowski TG, Worrell E (2011) Material efficiency: a white paper. Resour Conserv Recycl 55(3):362–381. https://doi.org/10.1016/j.resconrec.2010.11.002
9. Amudjie J, Agyekum K, Adinyira E, Amos-Abanyie S, Botchway EA (2023). Implementing the principles of circular economy in the construction industry: exploratory and confirmatory factor analyses of strategies. Constr Innov. https://doi.org/10.1108/CI-10-2022-0270
10. Bakker C, Hollander MD (2013) Six design strategies for longer lasting products in circular economy. The Guardian 10

11. Belagraa L, Kessal O, Boulaouad A, Cherif Mecheri M, Noui A, Abderrazak B (2020) Experimental investigation on the properties of a recycled aggregate concrete based on waste of the industrial mineral additions. KnE Engineering. https://doi.org/10.18502/keg.v5i4.6803
12. Benevolo C, Bovea MD, Preziosi M (2017) Sustainable refurbishment: a review of decision-making approaches and tools. Renew Sustain Energy Rev 76:92–110
13. Bilitewski B (2012) The circular economy and its risks. Waste Manage 32(1):1–2
14. Blengini GA, Garbarino E, Bevilacqua P (2017) Sustainability and integration between mineral resources and CDW management: overview of key issues towards a resource-efficient Europe. Environ Eng Manage J 16(2):493–502. https://doi.org/10.30638/eemj.2017.049
15. Bocken NM, Short SW (2021) Unsustainable business models–recognising and resolving institutionalised social and environmental harm. J Clean Prod 312:127828
16. Bocken NM, De Pauw I, Bakker C, Van Der Grinten B (2016) Product design and business model strategies for a circular economy. J Industrial Prod Eng 33(5):308–320
17. Bornstein D (2012) Bridging the clothing divide. Retrieved 1 May 2016 from http://opinionator.blogs.nytimes.com/2012/10/03/clothing-the-poorest-for-survival/?h=B&_r=1
18. Cachim P, Velosa AL, Ferraz E (2014) Substitution materials for sustainable concrete production in Portugal. KSCE J Civ Eng 18:60–66
19. Çimen Ö (2021) Construction and built environment in circular economy: a comprehensive literature review. J Clean Prod 305:127180
20. CIRIA (2011) Waste minimization in construction. CIRIA Special Publication
21. Den Hollander MC, Bakker C (2012) A business model framework for product life extension. In: Proceedings of sustainable innovation. Resource efficiency, innovation and lifestyles, pp 110–118
22. Desmond P, Asamba M (2019) Accelerating the transition to a circular economy in Africa. In: The circular economy and the global south. Routledge
23. Devenes J, Brütting J, Küpfer C, Bastien-Masse M, Fivet C (2022) Re: Crete–Reuse of concrete blocks from cast-in-place building to arch footbridge. Structures 43:1854–1867
24. de Andrade Salgado F, de Andrade Silva F (2022) Recycled aggregates from construction and demolition waste towards an application on structural concrete: a review. J Build Eng 52:104452
25. Eberhardt LCM, Rønholt J, Birkved M, Birgisdottir H (2021) Circular Economy potential within the building stock-Mapping the embodied greenhouse gas emissions of four Danish examples. J Build Eng 33:101845
26. Ecoinnovators (2015) The repair workshops—giving light to the age old art of repair. http://ecoinnovators.blogspot.com.es/2011/08/repair-workshops-giving-light-toage.html
27. EMF (Ellen MacArthur Foundation) (2013) Towards the circular economy: economic and business rationale for an accelerated transition. Ellen MacArthur Foundation. Accessed 5 Nov 2023. http://www.ellenmacarthurfoundation.org/business/reports/ce2012
28. Ellen MacArthur Foundation (2015) Growth within: a circular economy vision for a competitive Europe. Ellen MacArthur Foundation 100:81–92
29. Ellen MacArthur Foundation (2017) Case studies. https://archive.ellenmacarthurfoundation.org/case-studies
30. Ellen MacArthur Foundation (2020) What is the circular economy. https://www.ellenmacarthurfoundation.org/circular-economy/what-is-the-circular-economy
31. Esa MR, Halog A, Rigamonti L (2017) Developing strategies for managing construction and demolition wastes in Malaysia based on the concept of circular economy. J Mater Cycles Waste Manage 19:1144–1154
32. Esposito M, Tse T, Soufani K (2015) Is the circular economy a newfast-expanding market? Thunderbird Int Bus Rev. Retrieved from https://doi.org/10.1002/tie.21764
33. European Commission (2019) Circular economy implementation of the circular economy action plan. https://ec.europa.eu/environment/circular-economy
34. Evans J (2013) Circular economy toolkit. http://circulareconomytoolkit.org/about.html
35. Fastcompany (2014) Retrieved 1 May 2016, from http://www.fastcompany.com/most-innovative-companies/2014/industry/india

36. Fatemi S, Imaninasab R (2016) Performance evaluation of recycled asphalt mixtures by construction and demolition waste materials. Constr Build Mater 120:450–456
37. Ferreira J, Pinheiro MD, De Brito J (2015) Economic and environmental savings of structural buildings refurbishment with demolition and reconstruction—a Portuguese benchmarking. J Build Eng 3:114–126. https://doi.org/10.1016/j.jobe.2015.07.001
38. Fernández I, Kekäle T (2005) The influence of modularity and industry clockspeed on reverse logistics strategy: implications for the purchasing function. J Purch Supply Manag 11(4):193–205
39. Ferronato N, Rada EC, Gorritty Portillo MA, Cioca LI, Ragazzi M, Torretta V (2019) Introduction of the circular economy within developing regions: a comparative analysis of advantages and opportunities for waste valorization. J Environ Manage 230:366–378. https://doi.org/10.1016/j.jenvman.2018.09.095
40. Fitchett A, Rambuwani P (2022) Waste control in South African construction projects. Trans R Soc South Africa 77(2):105–112
41. Fonseca LM, Domingues JP, Pereira MT, Martins FF, Zimon D (2018) Assessment of circular economy within Portuguese organizations. Sustainability 10(7):2521
42. Francis D, Bessant J, Hobday M (2003) Managing radical organisational transformation. Manage Decis 41(1):18–31. https://doi.org/10.1108/00251740310462023
43. Frempong-Jnr EY, Ametepey SO, Cobbina JE (2023) Impact of stakeholder management on efficient construction waste management. Smart Sustain Built Environ 12(3):607–634. https://doi.org/10.1108/SASBE-08-2021-0147
44. Gamil A (2023) The impact of circular economy on economic development in Egypt. MSA Manage Sci J 2(2):39–79. https://doi.org/10.21608/msamsj.2023.194291.1007
45. Genovese A, Acquaye AA, Figueroa A, Koh SL (2017) Sustainable supply chain management and the transition towards a circular economy: evidence and some applications. Omega 66:344–357
46. Gerholdt J (2015) The 5 business models that put the circular economy to work. Retrieved 30 April 2016
47. Ghisellini P, Cialani C, Ulgiati S (2016) A review on circular economy: the expected transition to a balanced interplay of environmental and economic systems. J Clean Prod 114:11–32
48. Ghisellini P, Ulgiati S (2020) Circular economy transition in Italy. Achievements, perspectives and constraints. J Clean Prod 243:118360
49. Ghisellini P, Ripa M, Ulgiati S (2018) Exploring environmental and economic costs and benefits of a circular economy approach to the construction and demolition sector. a literature review. J Clean Prod 178:618–643
50. Ginga CP, Ongpeng JMC, Daly Ma KM (2020) Circular economy on construction and demolition waste: a literature review on material recovery and production. Materials 13(13):2970. https://doi.org/10.3390/ma13132970
51. Goyal S, Esposito M, Kapoor A (2018) Circular economy business models in developing economies: lessons from India on reduce, recycle, and reuse paradigms. Thunderbird Int Bus Rev 60(5):729–740
52. Homrich AS, Galvão G, Abadia LG, Carvalho MM (2018) The circular economy umbrella: trends and gaps on integrating pathways. J Clean Prod 175:525–543
53. Hossain MU, Ng ST, Antwi-Afari P, Amor B (2020) Circular economy and the construction industry: existing trends, challenges and prospective framework for sustainable construction. Renew Sustain Energy Rev 130:109948
54. Huang B, Wang X, Kua H, Geng Y, Bleischwitz R, Ren J (2018) Construction and demolition waste management in China through the 3R principle. Resour Conserv Recycl 129:36–44
55. Hazen BT, Mollenkopf DA, Wang Y (2017) Remanufacturing for the circular economy: an examination of consumer switching behavior. Bus Strateg Environ 26(4):451–464
56. Jayal AD, Badurdeen F, Dillon OW Jr, Jawahir IS (2010) Sustainable manufacturing: modeling and optimization challenges at the product, process and system levels. CIRP J Manuf Sci Technol 2(3):144–152

57. Jena SK, Sarmah SP, Padhi SS (2018) Impact of government incentive on price competition of closed-loop supply chain systems. INFOR: Inf Syst Oper Res 56(2):192–224
58. Kibert CJ (2016) Sustainable construction: green building design and delivery. John Wiley & Sons
59. Kibert CJ, Thiele LP, Davis LS (2000) Deconstruction: a growing industry in sustainable waste management. Waste Manage Res 18(4):304–313
60. King AM, Burgess SC, Ijomah W, McMahon CA (2006) Reducing waste: repair, recondition, remanufacture or recycle? Sustain Dev 14(4):257–267
61. Korhonen J, Honkasalo A, Seppälä J (2018) Circular economy: the concept and its limitations. Ecol Econ 143:37–46
62. Kumar S, Raut RD, Nayal K, Kraus S, Yadav VS, Narkhede BE (2021) To identify Industry 4.0 and circular economy adoption barriers in the agriculture supply chain by using ISM-ANP. J Clean Prod 293:126023. https://doi.org/10.1016/j.jclepro.2021.126023
63. Kabirifar K, Mojtahedi M, Wang C, Tam VW (2020) Construction and demolition waste management contributing factors coupled with reduce, reuse, and recycle strategies for effective waste management: a review. J Clean Prod 263:121265
64. Kamaruddin H, Marwan MA (2021) Extended producer responsibility (EPR) in Malaysia-towards a sustainable waste management system. J Contemp Issues Bus Govern 27(1):4490–4497
65. Korhonen J, Nuur C, Feldmann A, Birkie SE (2018) Circular economy as an essentially contested concept. J Clean Prod 175:544–552
66. Li RYM, Du H (2015) Sustainable construction waste management in Australia: a motivation perspective, pp 1–30. https://doi.org/10.1007/978-3-319-12430-8_1
67. Li J (2009) Summary policy and legislative tools in promoting waste management under circular economic model in China. In: International workshop on 3R and waste management. Kyoto City
68. Lieder M, Rashid A (2016) Towards circular economy implementation: a comprehensive review in context of manufacturing industry. J Clean Prod 115:36–51
69. Lund RT, Hauser WM (2010) Remanufacturing—an American perspective. In: 5th International conference on responsive manufacturing—green manufacturing (ICRM 2010), pp 1–6. https://doi.org/10.1049/cp.2010.0404
70. Loomba AP, Nakashima K (2012) Enhancing value in reverse supply chains by sorting before product recovery. Prod Plann Control 23(2–3):205–215
71. Maheshi D, Steven VP, Karel VA (2015) Environmental and economic assessment of 'open waste dump' mining in Sri Lanka. Resour Conserv Recycl 102:67–79. https://doi.org/10.1016/j.resconrec.2015.07.004
72. Mhlanga J, Haupt TC, Loggia C (2022) Shaping circular economy in the built environment in Africa. A bibliometric analysis. J Eng Des Technol. https://doi.org/10.1108/JEDT -03–2022–0175
73. Michelini G, Moraes RN, Cunha RN, Costa JM, Ometto AR (2017) From linear to circular economy: PSS conducting the transition. Procedia Cirp 64:2–6
74. Nguyen TC, Fishman T, Miatto A, Tanikawa H (2019) Estimating the material stock of roads: the Vietnamese case study. J Ind Ecol 23(3):663–673. https://doi.org/10.1111/jiec.12773
75. Ofori SA, Gyadu-Asiedu W, Assah-Kissiedu M (2015) Sustainable construction implementation in Ghana: focusing on awareness and challenges. Civ Environ Res 7(2):109–120
76. O'Grady T, Minunno R, Chong HY, Morrison GM (2021) Design for disassembly, deconstruction and resilience: a circular economy index for the built environment. Resour Conserv Recycl 175:105847
77. Ogunmakinde OE, Egbelakin T, Sher W (2022) Contributions of the circular economy to the UN sustainable development goals through sustainable construction. Resour Conserv Recycl 178:106023. https://doi.org/10.1016/j.resconrec.2021.106023
78. Ogunmakinde O, Sher W, Maund K (2019) An assessment of material waste disposal methods in the Nigerian construction industry. Recycling 4(1):13. https://doi.org/10.3390/recycling4010013

79. Oke A, Aghimien D, Aigbavboa C, Musenga C (2019) Drivers of sustainable construction practices in the Zambian construction industry. Energy Procedia 158:3246–3252. https://doi.org/10.1016/j.egypro.2019.01.995
80. Otasowie OK, Aigbavboa CO, Oke AE, Adekunle P (2024) Mapping out focus for circular economy business models (CEBMs) research in construction sector studies—a bibliometric approach. J Eng Des Technol. https://doi.org/10.1108/JEDT-10-2023-0444
81. Otasowie K, Aigbavboa C, Oke A, Adekunle P, Ayorinde E (2024) Diverse approaches to construction and demolition waste reuse: a case of South Africa. Eng Proc 76(1):2
82. Oyenuga A (2016) Economic and environmental impact assessment of construction and demolition waste recycling and reuse using LCA and MCDA management tools. https://www.academia.edu/64150567/economic_and_environmental_impact_assessment_of_construction_and_demolition_waste_recycling_and_reuse_using_lca_and_mcda_management_tools
83. Papargyropoulou E, Colenbrander S, Sudmant AH, Gouldson A, Tin LC (2015) The economic case for low carbon waste management in rapidly growing cities in the developing world: the case of Palembang, Indonesia. J Environ Manage 163:11–19. https://doi.org/10.1016/j.jenvman.2015.08.001
84. Park J, Tucker R (2017) Overcoming barriers to the reuse of construction waste material in Australia: a review of the literature. Int J Constr Manag 17(3):228–237
85. PBL Netherlands Environmental Assessment Agency (2018) Towards a safe operating space for the Netherlands. https://www.pbl.nl/sites/default/files/downloads/Towards_a_safe_operating_space_for_the_Netherlands_-_3333.pdf
86. Piccardo C, Hughes M (2022) Design strategies to increase the reuse of wood materials in buildings: lessons from architectural practice. J Clean Prod 368:133083
87. Pickin J, Randell P, Trinh J, Grant B (2018) National waste report 2018. Department of the Environment and Energy, Melbourne, Victoria, Australia
88. Pratt K, Lenaghan M (2015) The carbon impacts of the circular economy. Summary Report, June
89. Rathore P, Kota S, Chakrabarti A (2011) Sustainability through remanufacturing in India: a case study on mobile handsets. J Clean Prod 19(15):1709–1722. https://doi.org/10.1016/j.jclepro.2011.06.016
90. Ravindra K, Kaur K, Mor S (2015) System analysis of municipal solid waste management in Chandigarh and minimization practices for cleaner emissions. J Clean Prod 89:251–256. https://doi.org/10.1016/j.jclepro.2014.10.036
91. Reike D, Vermeulen WJ, Witjes S (2018) The circular economy: new or refurbished as CE 3.0?—exploring controversies in the conceptualization of the circular economy through a focus on history and resource value retention options. Resources, Conserv Recycl 135:246–264
92. Rijksoverheid N (2016) Nederland circular in 2050
93. Rios FC, Chong WK, Grau D (2015) Design for disassembly and deconstruction—challenges and opportunities. Proc Eng 118:1296–1304. https://doi.org/10.1016/j.proeng.2015.08.485
94. Røine K, Brattebø H (2017) Towards a methodology for assessing effectiveness of recovery systems: a process system approach. In: Perspectives on industrial ecology. Routledge, pp 101–111
95. Sabai MM (2015) Construction and demolition waste recycling into innovative building materials for sustainable construction in Tanzania. Am Sci Res J Eng Technol Sci (ASRJETS) 14(3):119–129
96. Sanches R, Mercan O, Roberts B (2018) Experimental investigations of vertical posttensioned connection for modular steel structures. Eng Struct 175:776–789. https://doi.org/10.1016/j.engstruct.2018.08.049
97. Schroeder P, Anggraeni K, Weber U (2019) The relevance of circular economy practices to the sustainable development goals. J Ind Ecol 23(1):77–95
98. Sholanke A, Aina-Badejo T, Aina-Babajide A, Nara Jacob A (2019) Noise pollution and waste control techniques in building construction in Nigeria: a literature review. IOP Conf Ser: Earth Environ Sci 331(1):012016. https://doi.org/10.1088/1755-1315/331/1/012016

99. Sihvonen S, Ritola T (2015) Conceptualizing ReX for aggregating end-of-life strategies in product development. Procedia Cirp 29:639–644
100. Simmonds PL (1862) Waste products and undeveloped substances: or hints for enterprise in neglected fields (Hardwicke R, Ed.)
101. Singh MP, Chakraborty A, Roy M (2018) Developing an extended theory of planned behaviour model to explore circular economy readiness in manufacturing MSMEs, India. Resour Conserv Recycl 135:313–322. https://doi.org/10.1016/j.resconrec.2017.07.015
102. Skaf M, Pasquini E, Revilla-Cuesta V, Ortega-López V (2019) Performance and durability of porous asphalt mixtures manufactured exclusively with electric steel slags. Materials 12(20):3306. https://doi.org/10.3390/ma12203306
103. Srivastava SK (2008) Network design for reverse logistics. Omega 36(4):535–548
104. Stahel WR (2010) The performance economy, 2nd ed. Palgrave Macmillan
105. Tang S, Wang W, Zhou G (2020) Remanufacturing in a competitive market: a closed loop supply chain in a Stackelberg game framework. Expert Syst Appl 161:113655. https://doi.org/10.1016/j.eswa.2020.113655
106. Tanskanen P, Sorvari J, Kurnitski J (2016) Refurbishment versus demolition and new construction—a study of the environmental impacts and costs of two residential buildings in Finland. J Clean Prod 139:555–564
107. Tse T, Esposito M, Soufani K (2015) Why the circular economy matters. European Business Review. Retrieved 3 Nov 2023 from http://www.europeanbusinessreview.com/?p=8372
108. Tse T, Esposito M, Soufani K (2016) How businesses can support a circular economy. Harvard Business Review. Retrieved 3 Nov 2023 from https://hbr.org/2016/02/how-businesses-can-support-a-circular-economy
109. Umar UA, Shafiq N, Ahmad FA (2021) A case study on the effective implementation of the reuse and recycling of construction, demolition waste management practices in Malaysia. Ain Shams Eng J 12(1):283–291. https://doi.org/10.1016/j.asej.2020.07.005
110. U.S. Environmental Protection Agency (EPA) (2016) Guide to industrial waste management, March. https://www.epa.gov/sites/production/files/2016-03/documents/industrial-waste-guide.pdf. Accessed on 5 June 2023
111. United States Environmental Protection Agency (2018) Reduce, reuse, and recycle construction and demolition materials at land revitalization projects
112. United States Environmental Protection Agency (2019) Advanced sustainable material management 2017 fact sheet. Accessed April 16, 2024 from https://www.epa.gov/sites/production/files/2019-11/documents/2017_facts_and_figures_fact_sheet_final.pdf
113. van Stijn A, Gruis VH (2019) Circular housing retrofit strategies and solutions: towards modular, mass-customised and 'cyclable' retrofit products. IOP Conf Ser: Earth Environ Sci 290(1):012035. IOP Publishing
114. Weimann K, Adam C, Buchert M, Sutter J (2021) Environmental evaluation of gypsum plasterboard recycling. Minerals 11(2):101
115. Wilhite DA, Sivakumar MV, Pulwarty R (2014) Managing drought risk in a changing climate: the role of national drought policy. Weather Clim Extremes 3:4–13
116. Worrell E, Reuter MA (eds) (2014) Handbook of recycling: state-of-the-art for practitioners, analysts, and scientists. Newnes
117. World Economic Forum (2016) Shaping the future of construction a breakthrough in mindset and technology. https://www3.weforum.org/docs/WEF_Shaping_the_Future_of_Construction_full_report__.pdf
118. Wright D, Meadows DH (2008) Thinking in systems. Earthscan
119. Xiao H, Li Z, Jia X, Ren J (2020) Waste to energy in a circular economy approach for better sustainability: a comprehensive review and SWOT analysis. Waste-to-energy:23–43
120. WRAP (2013) The true cost of food waste within hospitality and food service
121. Worrell E, Price L, Martin N, Hendriks C, Meida LO (2001) Carbon dioxide emissions from the global cement industry. Annu Rev Energy Env 26(1):303–329. https://doi.org/10.1146/annurev.energy.26.1.303

122. Yan J, Feng C (2014) Sustainable design-oriented product modularity combined with 6R concept: a case study of rotor laboratory bench. Clean Technol Environ Policy 16(1):95–109. https://doi.org/10.1007/s10098-013-0597-3
123. Yong R (2007) The circular economy in China. J Mater Cycles Waste Manage 9:121–129

Chapter 3
Theoretical Perspectives on Circular Economy in Organisations and Policies

Abstract The field of CE research has experienced significant growth and development, marked by an increasing recognition of the urgent need for sustainable and resource-efficient economic models. The overview of CE research encompasses a diverse range of studies, from theoretical frameworks to practical approaches, each contributing to a comprehensive understanding of how circularity can be integrated into economic systems. Various theoretical frameworks have been proposed to guide CE studies, emphasising closed-loop systems, product life extension, and resource optimisation. These frameworks provide a conceptual basis for exploring and implementing principles across different sectors and industries. Furthermore, CE policies translate theoretical concepts and research findings into actionable strategies. These policies take diverse forms, including regulatory measures, economic incentives, and collaborative initiatives. EPR, waste management targets, and circular design standards are among the many policy instruments that drive resource efficiency and reduce environmental impact. The objectives of CE policies are broad and interconnected. They encompass resource conservation, waste reduction, economic efficiency, job creation, innovation, and social equity. CE policies seek to create a more resilient and sustainable economic framework that aligns with global sustainability goals by addressing these objectives. The purpose of CE policies is to drive transformative changes in how resources are managed and utilised. These policies aim to shift economies from linear models to circular, regenerative systems. CE policies strive to achieve a harmonious balance between economic prosperity, environmental stewardship, and social well-being by promoting eco-friendly design, encouraging responsible consumption, and driving collaboration among stakeholders. Finally, the evolving landscape of CE research, theoretical frameworks, and practical approaches underscores the collective effort to transition towards more sustainable CE systems.

© The Author(s), under exclusive license to Springer Nature Switzerland AG 2025
O. Otasowie et al., *Circular Economy Business Model for Construction Organisations*, Green Energy and Technology, https://doi.org/10.1007/978-3-031-88322-4_3

3.1 Introduction

Theories examine the causes, mechanisms, and relationships behind actions, events, and ideas, offering a cohesive explanation or description of phenomena [40, 87, 193]. They play a crucial role in human inquiry by supporting the understanding, justification, characterisation, and prediction of occurrences. A range of theories has been proposed and continues to evolve to address the complex interconnections between sustainability and organisations, reflecting changes in how enterprises approach sustainability.

The concept of sustainable societies, which emerged in 1974 [58], laid the foundation for sustainable development. It gained widespread political attention through the Brundtland Report [174]. However, the definition of sustainability in the Brundtland Report has been criticised for its vagueness, adding to its complexity [173]. This ambiguity, coupled with its increasing importance in shaping national and international policies, has led to extensive debate and various definitions for sustainability-related concepts [101, 147]. Despite the variations, these concepts share a common emphasis on addressing sustainability challenges to ensure the survival and resilience of human systems by maintaining a balance between the ecosystem's carrying capacity and socio-economic systems.

To explore approaches for studying the circular economy (CE), several theories that provide a foundation for CE are examined. Furthermore, policy development and evaluation debates would lack focus and justification without a robust theoretical framework. Therefore, a thorough understanding of CE theory is crucial. Implementing a CE requires grasping the underlying theory and understanding its practical applications. CE theory provides an in-depth exploration of key aspects such as reuse, recycling, and the broader goals of CE. At its core, CE theory advocates for improved resource management to achieve both economic growth and environmental sustainability. Thus, a comprehensive understanding of CE is vital for driving a sustainable future while unlocking various business opportunities. Hence, this chapter examines CE perspectives as a sustainability concept and explores related policies.

3.2 Approaches on Circular Economy in Organisations

3.2.1 Corporate Social Responsibility

The concept of sustainability in relation to organisations can be traced back to the 1930s when early discussions on corporate social responsibility (CSR) began to surface in publications [30]. However, it was in 1953 that Howard Bowen's study, "*Social Responsibilities of the Businessman*", marked a significant step in theorising the connection between corporations and society [126]. Bowen argued that large corporations wield substantial power, and their actions and decisions profoundly influence the lives of individuals in various ways [24]. He defined the "*social responsibilities of businessmen*" as the duty to adopt policies, make decisions, and take actions aligned with societal values and objectives [24: 6].

Over time, the term "*corporate social responsibility*" gradually replaced the earlier phrasing of "*social responsibilities of businessmen.*" Despite this evolution, CSR faced significant criticism in the 1950s and 1960s. One of the most notable critiques came from Milton Friedman, who argued that a company's primary responsibility is to generate profits for its shareholders. According to Friedman, introducing broader social obligations risks undermining the free-market system's core principles [81]. CSR now encompasses an organisation's responsibilities toward society, the environment, and the economy. Its acceptance gained momentum in 1970 when the Committee for Economic Development (CED) released "*A New Rationale for Corporate Social Policy*" [126]. This publication highlighted how the sustainability of organisations is directly tied to the well-being of their surrounding environment. It argued that driving societal well-being aligns with the long-term interests of businesses, as a deteriorating environment could jeopardise the support systems on which they depend [14].

3.2.2 Stakeholder Theory

Stakeholder theory, originating in the late 1970s [137], offers a critical perspective on redefining the role of organisations. Freeman [79, 80] *Strategic Management—A Stakeholder Approach* introduced this framework, emphasising that businesses must understand their relationships with both traditional groups, such as customers, suppliers, and employees, and less conventional groups, such as environmental advocates, governments, and special interest groups [134]. Freeman [79, 80: 46] defined a stakeholder "as any group or individual who can affect or is affected by the achievement of the organisation's objectives." Stakeholders are typically classified into primary and secondary categories. The organisation directly or significantly influences primary stakeholders, whereas secondary stakeholders have a less immediate impact [31]. While the initial focus of stakeholder theory was on social stakeholders, it has increasingly incorporated non-social stakeholders to address environmental issues [139].

The stakeholder approach operates on the principle that businesses exist within a broader social environment and must respond to stakeholder pressures to achieve strategic goals [134]. This approach contrasts with the shareholder perspective, as Friedman [82] outlined, which prioritises profit maximisation for shareholders. In contrast, the stakeholder perspective argues that a company's primary responsibility is ensuring its long-term survival by balancing the interests of diverse stakeholders [97]. Beyond shareholders, these stakeholders may include employees, customers, governments, and the public, all of whom can significantly influence the organisation's success [126].

3.2.3 Corporate Sustainability

The term "corporate sustainability" (CS) has gained increasing prominence since the publication of the Brundtland Report [192]. However, a universally accepted definition of CS remains elusive. van Marrewijk [203: 107] describe CS as "demonstrating the inclusion of social and environmental concerns in business operations and interactions with stakeholders." Similarly, Artiach et al. [8: 31] define it as "a business and investment strategy that seeks to use the best business practices to meet and balance the needs of current and future stakeholders." A widely recognised framework for implementing CS is the triple bottom line (TBL), introduced by Elkington [62]. TBL expands traditional accounting, focusing solely on profit, by adding two dimensions: social and environmental performance. These "bottom lines" represent new metrics for evaluating business impact beyond financial outcomes. Slaper and Hall [187] noted that this approach differs significantly from conventional reporting systems. The adoption of TBL has grown over time, with scholars like Jamali [110] arguing that integrating TBL into management practices can drive sustainable growth. Markley and Davis [142] also suggest that emphasising TBL can enhance a company's competitive advantage. Beyond the TBL framework, sustainable business models (SBMs) offer another critical lens for understanding CS. SBMs incorporate TBL principles while addressing the interests of various stakeholders. Business models (BMs) typically explain how organisations create and maintain a competitive edge through their products or services. By integrating sustainability within their core strategies and operations, SBMs enable companies to align their goals with broader sustainability objectives, ultimately enhancing their competitiveness [18].

3.2.4 Green Economics

The green economy is essential for sustainable policy development [84]. Green growth emerged in intergovernmental discussions during the Fifth Ministerial Conference on Environment and Development (MCED) for Asia and the Pacific in 2005. Ever since, both the green economy and green growth have gained recognition as critical components of public policy at national and international levels. According to the United Nations Environment Programme [198: 5], a green economy seeks "to improve human well-being and social fairness while considerably lowering environmental dangers and ecological scarcities". Green growth is considered a cornerstone for achieving a green economy within sustainable development and poverty alleviation, as highlighted by the United Nations Economic and Social Commission for Asia and the Pacific [197]. This approach integrates economic progress with environmental stewardship, emphasising the need to transition production systems to adopt resource-efficient and environmentally sustainable methods [111].

Numerous policy frameworks and guides support the implementation of a green economy. Examples include the Europe 2020 strategy outlined by the European

Commission [66] and the Organisation for Economic Co-operation and Development's Towards Green Growth report [157]. However, Borel-Saladin and Turok [22] caution against substituting sustainable development with the green economy, instead, the two should complement each other. A genuinely sustainable economic model would prioritise resource efficiency and environmental sustainability by reflecting the full value of ecosystem services in market mechanisms. Achieving this requires market-based tools, legislative frameworks, technological innovation programs, and voluntary initiatives.

3.2.5 Co-evolution Theory

The co-evolution theory, grounded in Darwin's principles of natural selection [51], has often been employed to explain the dynamic relationship between businesses and their environment. Co-evolution refers to the mutual influence on the development of two or more species, where changes in one species drive evolutionary responses in the other. For instance, an adaptation in a plant's structure might benefit a specific herbivore species, which gains an advantage and becomes more prevalent through natural selection. This increased population then impacts the plant's evolutionary path, creating a cycle of mutual adaptation. Co-evolution emphasises that while natural selection operates independently, the interplay between species drives traits that are advantageous both collectively and individually.

In the context of sustainable development, co-evolution provides a valuable framework for examining how businesses interact with their surroundings. Early research loosely applied this concept to explore how organisations adapt to their environments [121, 130]. More recent studies have expanded on this foundation, exploring co-evolution in inter-organisational relationships and the broader environment [176]. Since the external environment influences both organisations and their alliances and simultaneously affects them through their actions, their evolution is inherently co-evolutionary. According to Das and Teng [52: 732], co-evolution involves the "simultaneous growth of organisations, alliances, and the environment independently and interactively."

3.2.6 System Theory

The Earth system, like any complex system, consists of numerous interconnected and interdependent components, as well as their surrounding environments [155]. These systems could self-assemble and organise without external intervention. Over the past 3.4 billion years, the Earth system has undergone self-assembly, reorganisation, and restructuring cycles, demonstrating resilience despite catastrophic extinction events. With an inherent creative drive, the Earth system can recover and adapt without human intervention, planning, or understanding. Nature can heal itself independently without the necessity of human oversight. Complex systems exhibit several defining characteristics. They are inherently asymmetrical and nonlinear, meaning that cause-and-effect relationships are not straightforward. Folke et al. [77] noted that large adaptive systems could display chaotic behaviours, multiple states of equilibrium, and transitions between different regimes, some of which may be irreversible. This nonlinearity means these systems are susceptible to abrupt and rapid changes, often resulting in unforeseen consequences and unpredictable outcomes. Such dynamics are a natural feature of complex systems [6].

Another key trait of systems is emergence, where the properties of the system differ from those of its components [16]. This emergent behaviour cannot be fully understood through reductionist thinking. Instead, emergent qualities arise independently because of the system's underlying parts [15]. The interactions among components create a higher level of complexity, leading to behaviours and responses that cannot always be anticipated using reductionist methods. Systems are also inherently suboptimal at every organisational level. This suboptimality reflects the trade-offs required to address multiple challenges simultaneously. Optimising a single aspect of a system would render other levels dysfunctional. Farnsworth and Niklas [71] argued that as the complexity of challenges increases, solutions must become progressively less optimal for each challenge. This principle applies to natural and human-designed systems, where trade-offs are an inevitable design reality.

Feedback is another crucial element of systems, enabling self-organisation and dynamism. Feedback loops provide real-time input, transmitting information within and across organisational levels. This continuous exchange drives changes that affect the system's functionality at all levels. Feedback acts as the "electricity" powering the system and the adhesive binding it together. While the Earth system provides constant feedback, humans often fail to recognise it due to their increasing detachment from the natural environment. Hence, Commoner [41] explored the idea of circularity, emphasising the importance of bridging the gaps between governance, technology, the environment, and the economy. The study highlighted the interconnectedness between a sustainable economy, the natural world, and human society. The study argues that true sustainability can only be achieved within the boundaries of the Earth system, reinforcing the necessity of aligning human activities with the planet's ecological limits.

3.3 Perspectives on Circular Economy in Organisations

3.3.1 Neoclassical Perspectives

Neoclassical theory, often associated with neoliberalism, was developed as a progression of classical economic theory. The term, first introduced by Thorstein Veblen in 1900, describes how the supply-and-demand framework integrates subjective and objective value theories. Alfred Marshall significantly contributed to this theory by blending traditional ideas, which attributed a commodity's value to its production costs, with newer insights suggesting that value is also derived from personal utility [124]. This theory emphasises the importance of deregulation, privatisation, and liberalisation as essential strategies for driving economic growth, advocating their adoption by all nations [1]. It is grounded in the belief that a free market drives competition, encouraging organisations to engage in the global economy. Neoclassical theory identifies the efficient allocation and distribution of scarce resources to maximise individual utility as a central economic challenge, ultimately contributing to national welfare. The approach strongly supports market freedom, allowing governmental and non-governmental entities, including individuals and businesses, to play active roles in economic planning.

Neoclassical theory has proven to be a robust framework, significantly advancing the economic understanding of environmental issues. However, the presumed connection between neoclassical economics and ecological sustainability is no longer universally accepted. Advocates of this economic school argue that it can offer insights into achieving sustainability. Unlike traditional environmental policies that rely on government planning and regulation, neoclassical economics proposes a market-based approach. This strategy, sometimes referred to as enabling policy, leverages market forces to address environmental challenges [28].

As sustainability has become a central focus in environmental policy discussions in recent years, the relevance of neoclassical economics has grown. Its approach aims to conceptualise environmental assets such as clean air, biodiversity conservation, and climate stability as commodities that can be integrated into conventional microeconomic analysis. According to Jacobs [109], this process involves two main stages of policy development. The first stage seeks to define environmental protection objectives by addressing the question, "How much of the environment should be preserved?" Neoclassical economics answers this by advocating for protection up to the point where the costs of conservation exceed the associated benefits. This approach relies on cost-benefit analysis, wherein the environmental benefits are assigned monetary values and compared over time against conservation costs. The second stage focuses on identifying the most effective tools to achieve these objectives, asking, "What methods will best accomplish the goals?" The neoclassical perspective emphasises efficiency, suggesting that the optimal strategy minimises societal costs. This perspective supports using market-based instruments, such as taxes, fees, and tradeable pollution permits. These tools incentivise businesses and consumers to adopt environmentally responsible behaviours while ensuring cost-effective solutions to ecological problems [109].

A market-based approach forms the foundation of the neoclassical theory. In this view, environmental challenges are considered market failures, where markets fail to deliver socially optimal outcomes. Such failures arise because environmental factors rarely assign a market price [109]. Consequently, the environmental costs, often termed external or social costs, are borne by third parties. The initial step in addressing these issues involves quantifying these external costs. This is done by estimating their market value or "shadow price," which reflects the costs and benefits associated with environmental factors. Market value is often derived by assessing the average consumer's willingness to pay for certain benefits or avoid specific costs. This can be achieved by examining market prices for goods associated with environmental attributes, such as scenic properties or rural tourism, or through contingent valuation surveys that ask individuals how much they would pay for environmental outcomes. In either case, consumer preferences in the marketplace, rather than government mandates, determine the perceived value of the environment [162].

The next step is to integrate these external costs back into the market. This process, known as internalisation, involves imposing taxes, fees, or tradable permits on activities that cause environmental harm. By doing so, the costs are directly assigned to those responsible, encouraging behavioural changes. When higher costs or taxes are enforced, the resulting shifts in market behaviour aim to reduce environmental damage. Ideally, this process reduces environmental impacts to a level where the marginal costs of harm equal the marginal benefits derived from mitigating it, assuming that cost estimates and policy instruments are effective [164]. Within the neoclassical tradition, sustainability highlights the limitations of market mechanisms in achieving environmental sustainability. This perspective aligns with the notion of "weak sustainability," which emphasises maintaining total capital (both natural and human-made) over time [163]. While this approach acknowledges the role of natural capital, it does not prioritise environmental preservation, as it permits environmental degradation if increases in human-made capital offset the losses [204]. Weak sustainability has thus evolved into a focus on sustaining overall well-being over time, often with minimal consideration of environmental components [166].

3.3.2 Institutional Perspectives

The institutional theory addresses why certain organisations exhibit consistent behaviour patterns [4]. DiMaggio and Powell [54] and Alvesson and Spicer [4: 203] define institutions as "regulative, normative and cognitive structures and activities that provide constancy and value for social behaviour," encompassing rules, laws, professional and social conventions, ethics, and traditions. In the context of environmental sustainability, institutions are known to influence organisations through a process called isomorphism, which leads organisations in similar geographic regions or contexts (e.g., industries or markets) to adopt comparable behaviours in response to shared environmental factors or pressures [4]. According to this theory, institutions impose coercive, normative, and mimetic pressures on organisations, resulting in isomorphic (similar) behaviours [4, 45]. Coercive isomorphism arises from the demands of those controlling essential resources for an organisation. In contrast, normative isomorphism stems from professional norms and practices shaped by education, training programs, professional associations, and employee groups [4, 53]. Mimetic isomorphism occurs when organisations facing uncertainty emulate successful peers to mitigate risk. These forces collectively drive organisations to align their behaviours with external expectations, driving business uniformity. However, institutional theory acknowledges exceptions. Alvesson and Spicer [4] suggest that middle-status firms often comply with institutional pressures to maintain legitimacy, high-status firms leverage their reputation to resist such pressures, and low-status firms prioritise survival over compliance. Similarly, Cormier and Magnan [45] found limited evidence supporting the role of institutional pressures in compelling organisations to meet legal standards.

Institutional theory is fundamental in explaining the relationship between environmental sustainability and institutional behaviour. It suggests that institutional actors often push the boundaries of legal requirements, compelling businesses to align with established conventions, rules, and laws shaped by internal and external influences. This alignment frequently results in businesses adopting environmentally protective measures. Jennings and Zandbergen [113] were among the first to analyse sustainable strategies through the lens of institutional theory. The study highlights how institutional elements contribute to the emergence and continuity of specific organisational forms, an idea that traces back to Meyer and Rowan [148]. Scott [184] identified multiple branches of institutional theory, while DiMaggio and Powell [54] emphasised how institutional forces drive isomorphism within organisational systems. Institutional pressures (coercive, mimetic, and normative) are central to spreading organisational practices, as demonstrated by numerous studies. However, research in the construction industry has primarily examined the outcomes of implementing these practices, often overlooking the processes that establish their legitimacy or the role of social movements in driving them [26, 116, 117].

Kauppi and Hannibal [117] applied institutional theory to propose that coercive pressure, particularly from legislative requirements, can encourage organisations to adopt environmentally sustainable practices, such as CE. This reflects procedural isomorphism, where institutional forces shape environmental management approaches across industries [218]. The extent of institutional environmental impact significantly influences practices, including reducing the costs of goods sold relative to returns. As an external driver, regulatory pressure is crucial in motivating organisations to adopt sustainability measures [2, 33, 103]. Non-compliance with regulations often leads to substantial costs, including penalties, highlighting the role of regulation in driving environmental commitments [50, 103].

Furthermore, the implementation of collaborative management across organisational domains influences the operations of various fields and professions, often shaped by normative pressures [117]. Normative institutions, such as trade unions, professional associations, and accrediting bodies, establish norms and regulations regarded as appropriate. This alignment of norms with what is deemed suitable ties normative pressures to appropriateness [128]. For example, industry associations enhance their sector's reputation, while professional groups exchange best practices and standardised procedures among companies [117]. According to Kauppi and Hannibal [117], an organisation's participation in such associations can influence its perception within its industry. Normative pressures also align closely with environmental sustainability programs, encouraging compliance with specific standards. Prescribed training programs represent one avenue through which normative isomorphism exerts influence [118]. Employees' practices are shaped by their training [128]. In construction management, for example, organisations could collaborate with academic institutions offering courses on building practices and construction management to promote sustainability-related standards [172]. Professionals who evaluate environmental sustainability within the construction sector are anticipated to drive firms toward adopting eco-friendly project methodologies. By driving partnerships between education and organisations, environmental sustainability can be normalised within the industry [172].

Similarly, mimetic pressures emerge in response to uncertainty. When clear guidelines are lacking, organisations may adopt practices similar to those of their peers [117]. Targets for imitation are often based on organisational attributes, performance, or frequency of use [117]. Mimetic behaviour can arise through direct interactions or by emulating the practices of organisations with similar traits, even without formal relationships [118]. Research on environmental sustainability highlights the significant influence of industry peers on organisational environmental policies [172]. Businesses within shared marketing networks often replicate the structures and strategies of others, using them as benchmarks [117].

Lastly, coercive pressures stem from political authority and legal requirements. These external pressures compel organisations to adhere to environmental laws and regulations or participate in environmental management programs [47, 54, 138, 178, 217]. Studies [47, 132, 180] demonstrate that coercive pressures significantly impact organisational behaviour concerning environmental practices. Organisations engage in environmental programs to gain benefits or avoid penalties and potential bans for violating regulatory standards [180]. Hence, regulatory authorities may enforce compliance with environmental sustainability standards through economic incentives, such as voluntary programs that offer subsidies or concessions, and through command-and-control mechanisms, which involve mandatory regulations [47, 132, 138, 180]. From the construction industry‘s perspective, companies strive to comply with laws and regulations to maintain legitimacy. Regulatory bodies employ various strategies, including providing information, offering incentives, using persuasive measures, and implementing legislation to reduce the environmental impact of construction activities [47]. Certification programs, which outline sustainability criteria, often require green practices like CE, compelling organisations to adopt these practices to qualify [47].

3.3.3 Neo-Marxist Perspectives

To address and assess the issues presented in this book, it draws upon neo-Marxist theories. Marx's critique of capitalism and his recognition of the need for recycling led him to develop a broader vision of ecological sustainability. While he believed that ecological sustainability held little practical relevance for capitalist society due to its inability to engage in rational, long-term planning, he considered it essential for a future society of cooperative producers. By emphasising the importance of preserving the environment for future generations, Marx captured the essence of the modern concept of sustainable development as "development which meets the needs of the present without compromising the ability of future generations to meet their needs" [211: 43].

Furthermore, Marx's concept of the metabolic rift illustrates how capitalist society disconnects people from the environment, despite critiques suggesting he overlooked ecological limits in production. He emphasised that metabolism is foundational for sustaining life, growth, and reproduction. Marx argued that urban pollution directly results from the failure to recycle materials back into the environment. He believed that, just as human metabolism produces waste that should be reincorporated into the natural cycle, industrial waste should be recycled into production alongside consumption and manufacturing waste [143]. According to Marx, capitalism disrupts the metabolic relationship between humans and nature by preventing consumed materials from returning to the environment. As consumption patterns grow, society focuses more on the products it creates and forgets the environmental resources that enable its production. Capitalism prioritises the production process over the source of raw materials, leading to a growing detachment from nature. This detachment drives the exploitation of natural resources to satisfy the demands of civilisation, deepening the rift between humanity and the environment.

In addition, neo-Marxist theories like the treadmill of production [182] offer insight into capitalism's environmental impact. The treadmill of production theory suggests that capitalism is characterised by an unending production cycle [108]. Schnaiberg argued that while natural energy flows are vital to human societies, altering these flows causes their degradation [13]. The treadmill of production involves two processes: additions and withdrawals. Additions refer to pollutants or waste returned to the environment, while withdrawals are raw materials extracted from nature to be converted into commercial products. Consumer and marketing-driven capitalism continuously pushes the production cycle to meet demand, accelerating resource depletion and increasing waste output [13, 92]. Following the laws of supply and demand, as demand rises, supply must increase, hastening production and its adverse environmental effects. Treadmill theory predicts that this unsustainable pattern will lead to environmental degradation, as human demands are infinite. At the same time, the environment's resources are limited, making the treadmill of production an untenable system [107].

Marx highlighted the fundamental contradiction within capitalism, specifically the exploitation of workers and the degradation of the environment. However, social structures are deeply entrenched in capitalist systems [13, 92, 107], with the primary goal of reinforcing and sustaining capitalist ideologies. For example, governments prioritise economic growth over environmental preservation, and labour unions support industrial expansion because it provides jobs [107]. As a result, the production cycle is unlikely to cease unless there is a significant shift in the global political and economic framework. Critics of this view argue that it places too much emphasis on materialism and believes that capitalism can adapt to address environmental issues without requiring drastic changes [92].

Immanuel Wallerstein's global systems theory, developed as a response to modernisation theory, expands on Marxist ideas. It argues that colonialism is a central factor in the underdevelopment of the Global South. Wallerstein shared Marx's view that capitalism seeks to maximise profit through exploitation. However, unlike Marx's focus on nation-states, Wallerstein's theory posits that capitalism operates globally, leading to a neo-Marxist perspective. He describes a global system comprising three regions: the core, the periphery, and the semi-periphery. The periphery provides labour and surplus resources to the core, while the semi-periphery both exploits and is exploited [207]. This creates a hierarchical global structure driven by commodification and rationalisation, which are key to the exploitative dynamics. An example of this process is the commodification of land. Building on Wallerstein's ideas, another theory on uneven ecological exchange suggests that environmental degradation and disasters arise from the environmental costs of high-consumption industrialised nations being shifted to less developed countries [114].

In conclusion, the neo-Marxist perspective offers a critical framework for understanding contemporary environmental challenges. It emphasises the need for profound changes in economic and social systems to achieve genuine sustainability. These views contribute to broader conversations about addressing environmental crises while tackling the root causes of inequality and exploitation.

3.3.4 Economic Perspectives

The CE as a sustainability concept from an economic perspective is centred on the trade-off between present and future consumption. For over two centuries, economists have explored sustainability in various forms. During the late eighteenth century, two prominent figures, Thomas Malthus and Adam Smith, viewed the same world through very different lenses. While Adam Smith's *The Wealth of Nations* [188] extolled the virtues of markets and the "invisible hand" as mechanisms for improving human well-being, Thomas Malthus foresaw a much bleaker future. Malthus argued that food production would grow linearly, whereas population growth would follow an exponential trajectory. This mismatch, he believed, would eventually exceed the land's capacity to sustain its population, leading to widespread famine and severe economic crises. In his view, economic expansion would inevitably trigger population surges, culminating in a collapse, resulting in a grim cycle of subsistence living. These projections earned economics the label of the "dismal science."

However, Malthus's predictions did not fully materialise, mainly due to the advent of the Industrial Revolution. Breakthroughs in agriculture and manufacturing drastically increased productivity, enabling the production of more goods and food with fewer resources. Such technological progress effectively negated the limitations that Malthus had envisioned, facilitating sustained economic growth. As constraints became less restrictive, economists turned their attention from the question of sustainability toward understanding how economic systems functioned in practice [65]. This shift in economic thought prevailed until the mid-twentieth century. By then, Boulding [23], in *The Economics of the Coming Spaceship Earth*, revisited Malthusian concerns. Boulding argued that in the context of a finite planet, some of Malthus's ideas were still relevant for addressing modern environmental and economic challenges. Boulding argued that traditional economic perspectives likened the world to an open, limitless frontier, an analogy drawn from the Old West, where resources seemed abundant, and expansion appeared unrestricted. While this view might have once held validity, Boulding noted a growing recognition that the planet operates more like a self-contained spacecraft, where resources are limited, and waste cannot be carelessly discarded. This "spaceship economy" requires frugality and careful resource management, emphasising the significance of both resource stocks and waste accumulation. Unlike the "cowboy economy," which focuses solely on resource flows, the spaceship economy prioritises sustainability by addressing the scarcity of critical resources and the long-term impacts of waste [23].

As Boulding highlighted, economic sustainability aligns with a renewed awareness of Malthusian concerns about resource limitations. The importance of resource access has increased in strategic economic considerations, but significant social and environmental consequences accompany it. Many resources extracted for economic purposes are not entirely consumed but become residual waste, posing risks to ecosystems and adding societal costs [10]. This recognition underscores the insufficiency of virgin supplies of certain critical minerals to sustain modern lifestyles globally without rethinking the current linear "take-make-dispose" economic model [90]. From an economic perspective, the CE offers distinct advantages, particularly resource efficiency. Building on long-standing waste and resource management approaches, CE emphasises the sustainable extraction, use, and disposal of materials. It provides individual, corporate, and governmental benefits by conserving the natural resources that support economic activities. Although resource efficiency and environmental protection efforts may sometimes appear to limit economic growth, businesses can often reduce production costs by optimising material inputs. This dual focus on sustainability and efficiency positions CE as a critical model for balancing economic and environmental goals.

These mutually beneficial scenarios could support the transition to a CE prioritising efficient resource use [202]. However, it is essential to distinguish between the short-term and long-term benefits of a resource-efficient CE to understand the concept of win–win outcomes fully. Over time, conserving the natural resources the economy heavily relies upon can yield significant benefits for economic sustainability. In the short term, businesses can gain from lower material input costs. By using fewer resources per output unit, organisations can reduce costs, enhance productivity, and improve competitiveness while mitigating risks associated with resource price fluctuations [202]. Furthermore, improving resource efficiency does not always guarantee immediate cost savings. Some enterprises may benefit financially from inefficiencies, such as overuse of resources and pollution because they do not directly bear environmental costs. Increasing resource input costs might be necessary to curb consumption. Such price adjustments could lead to a more equitable distribution of environmental costs related to resource extraction and processing, enhancing overall economic efficiency. While less efficient enterprises might incur short-term losses, these could be offset by gains in other sectors, ultimately resulting in net economic benefits [55, 100].

In addition, the macroeconomic impacts of resource efficiency are shaped by complex interactions between businesses, sectors, and industries. Economy-wide modelling serves as a critical tool for understanding these dynamics. This analysis highlights three key modelling studies linking resource efficiency to economic growth. First, the International Energy Agency [105] employed a computable general equilibrium (CGE) model to assess global energy efficiency initiatives, projecting a 0.4% increase in global GDP by 2035. However, the benefits are unevenly distributed, with energy-exporting nations experiencing reduced demand and lower prices. Second, the Ellen MacArthur Foundation [64] used a similar CGE model to explore potential CE advantages in Europe, estimating that improved efficiency in sectors like food, transport, and construction could contribute up to 12 percentage points to GDP growth by 2050. This growth is attributed to technological innovation but does not account for the costs associated with achieving these advancements, such as investments in training, research, and new production technologies. Lastly, CE and BioIS [32] used a macro-econometric model to analyse how tax reforms, private investments, and public spending could reduce raw material consumption in the EU28, highlighting the importance of policy and financial mechanisms in driving resource efficiency.

The findings suggest that increasing resource productivity by 2.0–2.5% annually can positively impact GDP. Beyond this threshold, the benefits of resource efficiency measures outweigh their associated costs. However, reinvesting revenues from environmental levies into the economy tends to have minimal or adverse effects on GDP. Instead, the primary driver of economic growth appears to be environmental tax reform (ETR) when paired with tax revenue recycling. Improved resource efficiency can also promote economic expansion, potentially increasing employment. However, of the three previously analysed economic models, only the CE and BioIS [32] model estimates employment benefits. According to this model, using proceeds from environmental levies to reduce labour taxes could stimulate job creation. A 2% annual improvement in resource productivity could result in nearly two million additional jobs by 2030. However, the model does not specify which industries, occupations, or regions will benefit from this job growth.

A shift toward labour-intensive service sectors offers significant potential for a resource-efficient CE. Conversely, jobs in resource-intensive sectors, such as mining, may decline. Policymakers must address these shifts, as such industries are often geographically concentrated. Environmental tax revenues could be utilised to manage these transitions, such as funding education and training programs to help workers adapt to new roles [158]. Also, enhanced resource efficiency can impact employment, as Walz [209] highlighted in an extensive analysis of five case studies. Two key insights emerge: First, industries with lower import dependency are likelier to experience net job growth, particularly in areas like product repair, which may otherwise contribute to exports. Second, sectors with higher labour intensity generally see more incredible employment growth. Hence, significant sectoral changes, especially from secondary to tertiary industries, underscore the need for education and training initiatives to facilitate workforce adaptation.

3.3.5 Social Perspectives

The Global Reporting Initiative [88: para 2] states that "the social dimension of sustainability concerns an organisation's impacts on the social systems within which it operates". Labuschagne et al. [122] expanded the understanding of social sustainability by incorporating a company's interactions with its stakeholders into its definition. Missimer et al. [150, 151: 42] adopted a systems-thinking perspective to explore social sustainability, identifying five core elements: "trust, common meaning, diversity, capacity for learning and capacity for self-organisation" as essential for meeting individual needs across a societal framework. CE principles, such as reuse, recyclability, and curbing resource consumption, could reshape the current economic model toward greater sustainability [19]. These approaches also extend the functional lifespan of products. Schröder et al. [183] highlighted the alignment between CE practices and the United Nations Sustainable Development Goals (SDGs) social dimension. A crucial social benefit of CE is its potential to create employment opportunities, addressing mismatches in skill sets and disparities in unemployment rates across different regions. Thus, governments must actively support CE-based enterprises by implementing new regulations and offering capitalisation incentives. Such policies would drive greener technologies, improve waste management, reduce carbon emissions, promote renewable energy adoption, and enhance overall energy efficiency. Creating green jobs would necessitate adapting business models to integrate innovations and respond to these advancements.

Pociovălişteanu et al. [167] explored the relationship between environmental policy and employment, identifying the rise of new job roles driven by such regulations. They emphasised the importance of education and training initiatives to prepare the workforce for these emerging roles and encourage stakeholders to adopt green technologies. Investments in environmentally sustainable infrastructure also play a crucial role in supporting a green economy. In this context, employment in a CE is pivotal for addressing socioeconomic challenges. CE can cater to the demands of a growing and evolving workforce by driving job creation and mitigating unemployment and social inequalities. Hence, governments and industries must collaborate to support this transformation by implementing policies and incentives to generate jobs to close material loops and enhance resource efficiency. In addition, the success of these efforts depends on equipping workers with the necessary skills and training to fulfil the demands of CE-related roles.

Furthermore, equity represents a vital dimension within the social perspective of CE. It plays a crucial role in driving social justice and protecting human rights. Naess [152] opines that equity's foundational principles include distributive justice, social justice, and the equity condition. These principles underline how social fairness aligns with sustainable development, which aims to address the needs of both current and future generations. While Geng and Doberstein [85] suggest that CE supports social fairness, Xue et al. [214] argue that CE contributes to more equitable social benefit distribution. In addition, Zhijun and Nailing [216] emphasise that CE can potentially drive social and economic progress, with some regions witnessing significant GDP growth following its implementation. This supports the argument that GDP growth could serve as an indicator of social justice. However, developing effective tools and processes is indispensable to fully integrating the concept of equity into CE.

Another critical social aspect of CE is the sharing or collaborative economy. Despite its widespread application, defining the sharing economy remains challenging due to its varying interpretations. Nevertheless, most models agree on its primary objective: enabling sharing resources, goods, or services among multiple individuals. Heinrichs [99] explains that the sharing economy is grounded in efficiently utilising resources facilitated by technology for lending, renting, or trading. Proponents like Cherry and Pidgeon [35] argue that this economy is supported by three key pillars: stronger social relationships (social), more sustainable resource use (environmental), and optimised financial resource utilisation (economic). Also, the sharing economy advances goals such as resource management, community development, economic empowerment, and creative expression. In recent years, the clarity of the sharing economy's definition has diminished as it increasingly overlaps with small business promotion and startup ecosystems. Technology platforms underpin and expand these collaborative models, which include product-service systems, redistribution marketplaces, and platforms for collaborative lifestyles. These models operate across various sectors, including retail, financial services, travel, lodging, transportation, and office space. However, while sharing models are integral to CE, they represent only a fraction of the broader framework. Thus, more thorough inquiries are necessary to understand the ramifications for users and clients and why it lacks adoption. To address these considerations, integrating digital technologies becomes essential for optimising resource utilisation and material flows. This also supports incorporating user and consumer perspectives into design processes, aligning them with circular solutions [27]. In addition, society can voice its opinions and influence decisions related to circularity through mechanisms such as participation and local democracy [20]. These approaches act as catalysts for neighbourhood transformation by informing and educating the public about the activities of relevant agencies and enabling them to engage in decision-making processes.

Furthermore, the involvement of society in such processes provides valuable insights and can contribute significantly to tackling persistent social challenges in an inclusive and multidimensional way. It can also drive innovation in circularity efforts. Consequently, CE strategies could include precise, methodical, and strategic approaches to drive stakeholder collaboration to achieve policy coherence. Critical steps include strengthening stakeholder connections, integrating diverse perspectives into policymaking, and participating in international dialogues. Hence, a shared understanding of a sustainable CE system should underpin these efforts. Korhonen et al. [120] highlight that enhanced community engagement can lead to practical benefits from a local perspective, encouraging more democratic decision-making processes when a CE system is in place. From a stakeholder perspective, ensuring such participation can improve the quality and feasibility of decisions, enhance legitimacy, and deepen awareness of the complexity of societal challenges. Local actors can be pivotal in advancing community-focused approaches, emphasising local empowerment and active involvement in CE initiatives.

Finally, transitioning to a CE has implications for human health priorities in the coming years. These priorities include tackling significant health challenges, improving public health systems, and driving resilient and supportive communities. Environmentally friendly shifts in CE practices, such as adopting sustainable transportation and manufacturing methods, can reduce air pollution, thereby benefiting public health [160, 179]. However, it is equally essential to mitigate risks associated with hazardous chemicals, as failing to manage these risks could exacerbate disease burdens [11]. Thus, by introducing cost-effective and efficient solutions, the CE can enhance public health and healthcare service delivery. Ultimately, transitioning to a CE can contribute to building resilient communities and ensure improved quality of life and well-being.

3.4 Circular Economy Policy

Policy can be understood as a structured framework of principles, guidelines, and actions created by governmental or organisational authorities to address specific issues, achieve objectives, and regulate behaviour [5]. Policies serve as essential tools for shaping societal, economic, and governance dynamics. Policies establish a roadmap for decisions and actions within an organisation, aligning them with the organisation's overarching goals, senior management's objectives, and core management philosophies. These policies act as foundational principles that provide direction, guidance, and influence over organisational decision-making. The term "guiding" is a central concept in policy-making, emphasising their role in offering strategic direction while allowing for operational flexibility [42]. Policies typically outline goals, objectives, and underlying management philosophies. Factors such as business strategy, market competition, legislative requirements, and other external pressures often influence policy formulation. At their core, policies are formal organisational decisions designed to support operations and articulate concepts that stakeholders are expected to follow. Each policy should address critical issues relevant to achieving an organisation's primary goals. For example, a workplace health and safety policy connects directly to strategic objectives like improved service quality, reduced costs, and minimised injuries, highlighting its importance to the business and its stakeholders. Noncompliance with such policies can result in penalties, legal consequences, reputational harm, or risks to employee safety, depending on the industry context.

When examining policies in the context of the CE, they can be categorised as initiatives led by governmental or organisational authorities to facilitate and regulate the transition from a linear to a circular model. Government involvement plays a pivotal role in enhancing environmental management practices within businesses. Effective policies, rules, and regulations are necessary to support adopting CE principles, particularly in emerging economies, where governmental intervention is critical for enabling the shift. Transitioning to a CE demands substantial adjustments in governmental regulations, corporate strategies, and consumer behaviours. Therefore, it is imperative to develop coherent and sector-specific policies that address these changes [141, 201]. Braun et al. [25] emphasise the importance of collaboration and shared understanding among all stakeholders in creating value within a CE framework. Governments can promote CE adoption by establishing tailored environmental standards for different industries [185] and driving public awareness of CE principles [145, 201]. To integrate CE principles into existing environmental laws and regulations, governments could design comprehensive systems that include financial incentives, technical support, collaboration platforms, transparent supply chain monitoring protocols, and precise auditing mechanisms [129]. These efforts are essential for driving the systemic change required to achieve CE.

Several studies [83, 115, 141, 201] emphasise the critical role governments and regional authorities play in shaping frameworks to guide, promote, and monitor the transition to CE. Governments can facilitate the shift to CE by providing necessary infrastructure and technology [85, 141] and implementing policies such as new regulations, eco-labels, and laws [169]. In addition, they can drive circularity through public procurement initiatives, legislative support, and the adoption of circular business practices [215]. For example, the European Commission has developed a CE roadmap with strict guidelines for managing and storing waste, such as old tiles, in line with EU waste management laws. Furthermore, some countries have imposed restrictions on importing secondhand goods, including textiles. At both the macro and national levels, government policies are rapidly evolving to support CE adoption [21]. In the Netherlands, the government introduced the Waste to Resources policy (Gr Van Afval Naar Grondstof) to encourage circular production models [127]. Hartley et al. [95] explored CE-related policy expectations through interviews with 47 EU experts. The study identified key issues such as raising awareness, facilitating free trade of waste, reducing taxes on compliant products, adhering to manufacturing standards within the EU, and extending the circular supply chain.

Governments may also support CE through eco-industrial parks and virtual platforms for waste reuse. An example of government-led intervention is the Chinese central government's establishing a national CE policy in 2008 [144]. Dubey et al. [60] identified governments as crucial external drivers for achieving sustainable supply chains. By promoting green procurement and supply initiatives, governments can encourage manufacturers to adopt CE practices [115]. Public procurement also has the potential to increase consumer demand for recycled products. Policies concerning material handling and government procurement are essential for implementing circular procurement models [140]. The government's role extends to serving as a coordinator, mediator, and facilitator among various stakeholders. This involves crafting economic policies, enacting regulations, raising public awareness, and driving collaboration across the supply chain [104]. Perry [165] highlighted the need for government agencies to collaborate and interact with stakeholders such as clients, retailers, suppliers, and manufacturers to integrate CE principles into supply chains. Also, governments contribute to multi-stakeholder initiatives by setting, coordinating, and enforcing standards across all levels of the supply chain. Tsoi [195] stressed the importance of collaboration among regional governments, non-governmental organisations (NGOs), and multinational corporations (MNCs) to achieve sustainable business practices.

3.4.1 Circular Economy Policy Framework

CE policy frameworks have advanced considerably in recent years as governments, businesses, and societies acknowledge the urgent need for sustainable and resource-efficient systems. Initially, these frameworks emerged in response to the environmental challenges inherent in linear economic models, which rely on resource extraction, use, and disposal. As highlighted in the Intergovernmental Panel on Climate Change (IPCC) 2022 report, human-driven climate change is profoundly altering the planet. Its effects are evident in the significant disruptions to freshwater, marine, and terrestrial ecosystems and its detrimental impact on human infrastructure and health [168]. The Anthropocene era, marked by substantial human influence on the environment, is believed to have begun in the late eighteenth century [46]. Scientific research shows that global temperatures have risen since 1950, with current warming at 1.01 °C above 1880 levels, and continued greenhouse gas emissions are expected to exacerbate this trend [154].

The linear economy, characterised by the extraction of raw materials, their conversion into products, and eventual disposal, is centred on maximising sales and profits. This model encourages the production of goods that meet basic functionality but are designed for short-term use, leading to frequent disposal [93]. While this approach fueled economic growth and material wealth during the industrial era, its limitations have become evident in the twenty-first century. EMF [63] reported that commodity prices, after peaking in 1999, experienced unforeseen volatility due to declining manufacturing costs, depletion of easily accessible extraction sites, and rising technical challenges in mining new locations. Increased demand further drove production costs upward, creating financial pressures on manufacturers and limiting their ability to pass on costs to consumers. Consequently, the linear economy's emphasis on resource extraction, production, and waste management has become increasingly unsustainable [86, 96]. Environmental issues, such as water scarcity and pollution, have become pressing global concerns. In the coming decades, access to clean water is projected to become even more constrained, exacerbated by industrial waste and emissions contaminating land, water, and air. Increased production and consumption have also intensified natural resource depletion and the release of harmful pollutants into the environment. Deforestation and processing of raw wood significantly contribute to environmental degradation by reducing vegetation cover, generating waste, and disrupting ecosystems. These activities decrease oxygen production, increase atmospheric carbon dioxide, and threaten wildlife habitats, resulting in the loss of numerous animal and bird species. To address these challenges, immediate action is required. Investments in carbon capture technologies, reductions in greenhouse gas emissions, and strategies to mitigate human environmental impacts are critical to driving a sustainable future.

Considering the above, waste management policies experienced a notable shift over the late 20th and early twenty-first centuries. Initially, waste management strategies focused on mitigating local environmental pollution by adopting ecologically sound practices. However, implementing and promoting 3R (reduce, reuse, recycle) principles marked a transition toward sustainability-centered policies. The adoption of significant legislative measures in the United States reflects this evolution. The Resource Conservation and Recovery Act (RCRA), extensively amended in 1984, laid the foundation for solid and hazardous waste management. While municipal solid waste (MSW) management remains under state jurisdiction, the federal framework supports resource conservation and pollution prevention. The Pollution Prevention Act, revised in 2002, further advanced this agenda by incorporating these principles into strategic planning efforts, such as the 2010–2014 Strategic Plan. This plan aimed to achieve goals like reducing greenhouse gas (GHG) emissions, limiting hazardous materials, conserving water, driving profitable enterprises, and integrating effective pollution prevention methods. The Environmental Protection Agency (EPA) highlighted the interconnected benefits of achieving these objectives [200]. Another noteworthy initiative, the Resource Conservation Challenge, complements the RCRA and the Pollution Prevention Act. This program focuses on pollution prevention, promoting recycling and reuse, reducing hazardous chemical usage, and conserving resources and energy. It sets specific MSW recycling targets and prioritises resource management to address GHG emissions and environmental conservation. Together, these efforts form an integrated waste management system in the U.S., aiming to manage hazardous substances, lower GHG emissions, and preserve natural resources.

In the European Union (EU), the Waste Framework Directive (2008) provides the overarching legal framework for waste management, with member states aligning their national policies accordingly [68]. The EU stands out for harmonising waste management regulations with 3R principles. Key directives illustrate this integration, such as the Incineration of Waste Directive (2000) and the Landfill of Waste Directive (1999) [69]. The Incineration Directive mandates Best Available Technology (BAT) and Best Environmental Practices (BEP) to curb dioxin emissions, while the Landfill Directive promotes a phased reduction of organic waste sent to landfills, driving the adoption of 3R policies [70, 76]. Additional EU legislation emphasises recycling and pollution reduction, including the 2002 directives on Waste Electrical and Electronic Equipment (WEEE) and the Restriction of Hazardous Substances (RoHS). WEEE regulations encourage establishing collection points and recycling electronic waste, while the RoHS Directive restricts hazardous substances in products. The principles of extended producer responsibility (EPR) underpin these initiatives, requiring manufacturers to manage their products' lifecycle, from design for recyclability to post-use recovery. Germany pioneered EPR for packaging waste in 1992, setting a precedent for further EU legislation on waste batteries and end-of-life vehicles [199].

Furthermore, the UK recognises the importance of Extended Producer Responsibility (EPR); however, its waste management strategy emphasises the principle of shared responsibility. This approach assumes that all stakeholders (producers, consumers, and others) share accountability for a product's lifecycle, from creation to disposal and recycling [98]. In line with this, the UK government and various organisations have voluntarily agreed to minimise waste and promote the recycling of materials such as paper, food, and packaging. In addition, a landfill tax was introduced, starting at £40 per tonne, with annual increments of £8 per tonne, reaching £72 per tonne by 2013. Economic measures such as a unit-based fee system in Italy encourage improved waste management practices. This system currently covers about 15% of the municipalities, representing 29% of the population. Fees are determined based on the amount of waste generated, aligning financial incentives with waste reduction goals [57].

Japan's waste management strategy is guided by the 2000 legislation for establishing a Material Cycles Society. This framework promotes the 3Rs (reduce, reuse, recycle) to conserve natural resources and reduce environmental impact [112]. Targets were set for metrics like resource productivity, recycling rates, and ultimate waste disposal volumes. These principles are further supported by global initiatives such as the G8's 3R framework. Japan's regulatory system includes two core laws: the Waste Disposal and Public Cleansing Law (amended in 2010), which addresses waste management, and the Law for the Promotion of Effective Utilization of Resources [125], which governs recycling. Specific rules apply to waste categories like food, construction materials, electrical equipment, packaging, and end-of-life vehicles (ELVs). Also, the 2001 Law on Promoting Green Purchasing encourages government procurement of recycled products, underscoring Japan's commitment to sustainable practices.

In South Korea, the Waste Management Act (revised in 2007) and the Act on Promotion of Resource Saving and Recycling, updated in 2008, provide the primary legislative framework for waste management [112, 149]. A comprehensive amendment 1991 integrated recycling, landfill regulations, and incineration protocols into the Waste Management Act [112]. The 2002 Act on Promotion of Resource Recycling, later incorporated into the 2008 legislation, established guidelines for material reuse, cost structures, packaging regulations, and EPR systems. Specific measures for electronic equipment and vehicles were introduced under the 2008 Act on Resource Recycling of Electrical and Electronic Equipment and Vehicles [161]. Grassroots campaigns opposing waste facility construction prompted the introduction of the 3R policies. In response, the 1995 Promotion of Installation of Waste Disposal Facilities and Assistance to Adjacent Areas Act was enacted to ensure proper facility placement. Furthermore, economic measures like waste generation charges and bans on single-use packaging were implemented. Since 2005, organic waste has been mandated to be recycled rather than landfilled [112].

Furthermore, China's foundational Environmental Protection Law, established in 1989, laid the groundwork for subsequent waste management policies. In 1996, the Solid Waste Environmental Pollution Prevention and Control Law was enacted and revised in 2005. This law promotes the application of the 3Rs to reduce waste quantity and hazard levels [37]. While the primary focus is on detoxifying waste, the legislation does not prioritise specific treatment methods such as landfilling or incineration. It also introduced a hazardous waste inventory to manage such materials better. China introduced its primary legislation on material cycles and waste management, the CE Promotion Law, in 2008 [186]. This law was implemented to address various challenges, including limited financial resources, low utilisation of recycled materials, and the need to tackle resource depletion while pursuing sustainable economic growth. China's rapid economic expansion increased resource consumption and significant waste generation, resulting in pressing social and environmental concerns. Consequently, upgrading the waste management infrastructure became imperative. The CE Promotion Law emphasises promoting zero emissions at production sites and the efficient use of recycled materials, particularly in industrial zones with high concentrations of facilities. It prioritises managing and encouraging 3R (Reduce, Reuse, Recycle) activities within the industrial sector. In addition, it offers financial incentives such as reduced or exempted value-added tax to support the circular and integrated use of resources [131]. To make the CE Promotion Law effective, improvements were required in recycling and waste treatment practices and strengthening businesses' financial base by utilising recycled materials. This was particularly essential for the effective management of WEEE and MSW. China's waste management policies have aimed to balance resource conservation with economic growth. However, a recognised need has been to enhance environmental safeguards for imported recycled materials and address the rising MSW caused by urbanisation and population growth. In response, China enacted the WEEE Management Ordinance (2009) and the ELV (End-of-Life Vehicles) Recycling Law (2001) to tackle these challenges [131]. The ELV Recycling Law focuses on preventing unauthorised vehicle reconstruction, ensuring safety, and encouraging the purchase of environmentally friendly models. However, it falls short of guaranteeing proper ELV waste disposal, and existing recycling facilities lack sufficient capacity to handle the projected increase in ELV volumes [34, 210].

In Vietnam, the Environmental Protection Law (revised in 2005) is the cornerstone of the country's environmental and waste management framework, replacing older legislation. The 2007 Decree on Solid Waste Management outlines strategies for solid waste handling, emphasising recovery, treatment, and recycling to minimise reliance on landfills [36]. The decree also details waste management processes and associated costs, with rates for municipal solid waste at 40,000 VND per ton and hazardous waste charges reaching up to 6,000,000 VND per ton [206]. It further categorises hazardous and medical waste, aiming to strengthen waste management practices nationwide. Vietnam's waste management objectives for 2025 and 2050 are outlined in the *National Strategy on Integrated Solid Waste Management* (SWM), established in 2009 [206]. A key target for 2025 is the separation of MSW. By this time, 90% of construction waste in urban areas and municipal waste in suburban regions should be collected. In addition, all hazardous and non-hazardous waste from industrial operations and solid waste from businesses in urban areas must be managed using environmentally sustainable practices. Looking toward 2050, the strategy emphasises minimising the volume of waste sent to landfills by implementing 3R (Reduce, Reuse, Recycle) initiatives, ensuring comprehensive solid waste collection, and adopting advanced environmentally friendly technologies. The scarcity of land suitable for landfills has intensified efforts to promote 3R practices, particularly in urban centres [205].

In South Africa, waste management has evolved from a fundamental approach focused on maintaining cleanliness through proper disposal to a more comprehensive system guided by the Waste Act 2008 (Act No. 59 of 2008). Since its implementation on July 1, 2009, the Act has mandated the creation of a National Waste Management Strategy (NWMS) to achieve several goals: driving a cleaner environment, reducing waste generation, and encouraging reuse, recycling, and treatment, with disposal as a last resort [191]. The policy highlights the incompatibility of poor waste management with the broader objective of harmonious coexistence and warns that such practices threaten the fundamental right to a healthy life.

When considering South Africa's waste management practices within a global framework, several aspects become apparent. As Africa's most industrialised nation [196], South Africa faces environmental challenges typical of more developed countries. The country adopts globally recognised best practices in its waste policies, enacted into legislation after careful review and promulgation. Recycling initiatives are in place, but their effectiveness is not always apparent. A notable concept within South Africa's waste strategy is transforming waste into economic value, encouraging public and private sector participation across the waste management value chain. This approach also presents opportunities for poverty alleviation and job creation [175].

3.5 Approaches to Circular Economy Policy

Implementing CE principles aims to address the shortcomings of traditional waste management by highlighting the benefits of innovative business models to decision-makers [177]. CE emphasises the value inherent in recycled materials, making the approach both practical and beneficial [72]. However, effective implementation requires the integration of diverse policy frameworks throughout the waste management value chain. These policies encompass a variety of strategies designed to enhance sustainability, improve resource efficiency, and reduce waste generation. In addition, such policies help overcome sociocultural, institutional, and economic barriers [181]. They also respond to the increasing environmental concerns of Millennials and Generation Z, who view CE as a moral imperative for addressing climate change while strengthening supply chain resilience [7]. The following sections discuss specific CE policy approaches.

3.5.1 Extended Producer Responsibility

Over the last two decades, several nations have implemented a growing number of regulations and policies to reduce the environmental impact of products. These efforts underscore the continued commitment of governments to address environmental challenges [156]. A significant foundation for these policies is the principle of Extended Producer Responsibility (EPR), which assigns manufacturers accountability for the environmental effects of their products throughout their life cycles [59, 78, 146, 153, 156]. Grounded in the polluter-pays principle (PPP), EPR emphasises addressing the entire life-cycle impact of products while integrating the costs of environmental externalities associated with end-of-life (EOL) goods. It also promotes environmentally conscious product design [74, 119, 156]. Traditionally, environmental issues were managed through legislative approaches relying on a "command and control" model, primarily targeting "end-of-pipe" pollution problems [12, 49]. However, EPR-based policies have shifted the focus from managing pollution at the process level to considering the environmental implications of products throughout their entire life cycle, from resource extraction to disposal, recycling, and reuse [73, 156].

The Organisation for Economic Co-operation and Development (OECD) defines EPR as an environmental policy framework that extends a producer's responsibilities to include the disposal of products after consumer use [159, 208, 212]. This approach places the financial and physical burden of recycling on upstream producers while encouraging environmentally responsible product design [159, 208, 212]. Both manufacturers and consumers contribute to waste generation, and the system also incentivises changes in consumer behaviour [146, 153]. EPR has been adopted using various administrative, economic, and informational policy tools [78, 123, 133]. Key policy instruments employed by OECD nations include economic tools and mandatory product take-back strategies [156, 208, 209]. Examples of these strategies are as follows:

1. **Mandatory product take-back and recycling targets**: These policies set specific recycling objectives and require producers or retailers to accept EOL products. Industries often form Producer Responsibility Organisations (PROs) to manage EPR obligations collectively [212].
2. **Voluntary take-back programs**: Producers voluntarily meet take-back and recycling targets without penalties for failing to achieve them.
3. **Mandatory take-back with tradable recycling credits**: In this system, manufacturers must meet take-back and recycling targets but can trade credits to comply with requirements.

Economic tools employed include:

1. **Advanced Recycling Fee (ARF)**: The ARF involves a tax levied on the sale of products to cover the costs of recycling goods at the end of their life cycle. This fee is calculated at the point of sale and can either be charged directly to consumers or upstream to producers, with the cost integrated into the product's retail price. The fee is generally applied on a per-unit basis.
2. **Recycling Fee with Subsidy**: This mechanism uses funds collected through ARF or post-consumption recycling charges to support recycling activities. These funds can be allocated to various needs, such as covering waste management costs, developing recycling infrastructure, or subsidising producers to encourage the recycling of waste materials.

3. **Deposit Refund System (DRS)**: The DRS combines a deposit paid at the time of purchase with a refund offered when end-of-life (EOL) products are returned for recycling or environmentally safe disposal. The deposit reflects the product's market value and the environmental costs of recycling. This system ensures a consistent flow of materials for recovery and recycling while incentivising resource reduction and reuse.

3.5.2 Upstream Tax Categories

Two main types of upstream taxes are commonly used to drive sustainable practices:

1. **Material Taxes**: These taxes are imposed on hazardous or difficult to recycle materials, encouraging manufacturers to use safer and more recyclable alternatives. The tax amount is determined based on the cost of managing the material versus the marginal cost of treatment.
2. **Combination Tax/Subsidy**: This approach involves a tax paid by producers, which is then used to fund waste treatment facilities [159].

Experience in South Africa highlights the potential of voluntary, industry-led EPR initiatives to achieve recycling targets without government enforcement. For instance, three voluntary programs have been implemented to address post-consumer waste. These programs focus on eliminating 100% recyclable products from the waste stream and creating economic opportunities for stakeholders [153]. One prominent example is Collect-a-Can [38, 39], a non-profit organisation established in 1993 as a joint initiative by Nampak (Africa's most prominent beverage can and packaging manufacturer) and ArcelorMittal South Africa (the leading supplier of steel and tinplate for food and beverage cans). These two companies fund this initiative. The primary goal of Collect-a-Can is to implement EPR measures on behalf of the industry while avoiding regulatory actions that could harm its interests. The program rewards collectors and consumers with payments exceeding market value for returning used cans. Although the steel industry does not require the tin content in the cans, ArcelorMittal has voluntarily incorporated them into its EPR commitments, using the cans with other waste materials to produce mild steel. This initiative provides collectors with a source of income and encourages entrepreneurship in the recycling sector by supporting individuals in starting their recycling businesses.

In 2004, the collaborative industry initiative PETCO was established to tackle the challenge of recycling polyethylene terephthalate (PET) bottles in a non-profit manner. PETCO runs EPR programs on behalf of its stakeholders in the PET sector, which include resin manufacturers, bottlers, converters (who create bottles from PET resin), and brand owners such as Coca-Cola. The initiative resulted in a Memorandum of Understanding (MoU) between the industry and the Department of Environmental Affairs and Tourism (DEAT), agreeing to postpone the enforcement of PET-related regulations in exchange for jointly established goals, evaluation methods, and monitoring systems. PETCO has introduced a system of advanced recycling fees and incentives, where converters and bottlers pay a voluntary fee for every tonne of PET resin purchased from importers and manufacturers. The funds raised cover operational costs, support recyclers and recycling programs, and assist businesses promoting PET recycling. To protect recyclers and collectors from market fluctuations, PETCO ensures the price of recycled PET artificially increases during adverse market conditions, thus maintaining their participation in the industry.

3.5.3 Incentives and Subsidies

Subsidies are financial aids the government provides to individuals, organisations, or corporations and can be either direct (such as cash transfers) or indirect (like tax breaks). These subsidies are generally designed to ease certain hardships and are often seen as being in the public interest, supporting economic policies or social benefits. Research has shown that subsidies can effectively drive environmental innovation in products [48], encourage recycling efforts [91], reduce environmental damage, and enhance social welfare [135]. As a result, several countries, including the USA, China, and Japan, have developed subsidy programs to support waste recycling initiatives [170]. For example, California imposed taxes on producers but provided subsidies to recyclers [94], and Japanese companies have developed battery recycling systems under national subsidy laws [9]. In Hunan Province, China, the Liuyang government offered a one-time subsidy covering 20% of the construction costs for remanufacturing [136]. Such initiatives generally focus on the final waste-handling phase or the recycling stage in the manufacturing-recycling system. Countries have passed laws urging manufacturers to adopt eco-design practices to promote waste reduction and pollution prevention. In 2009, the Canadian Council of Ministers of the Environment (CCME) introduced a strategy to encourage eco-design and maximise environmental benefits [29]. In addition, to promote the production of environmentally friendly products, the US government granted $2.4 billion to electric vehicle manufacturers [89].

Subsidies are a key tool used by governments to promote recycling and eco-design. These subsidies typically fall into performance-based (aimed at encouraging eco-design or increasing recycling rates) and cost-based (focused on funding eco-design or recycling investments). The effectiveness of these programs depends on their type and purpose. Eco-design is used as a measure of a manufacturer's performance in waste avoidance while recycling volume serves as an indicator of a recycler's performance in waste treatment. The better a manufacturer's eco-design, the higher their performance; similarly, recyclers improve their performance with more significant recycling volumes. As such, performance-based subsidies can be further divided into recycling quantity subsidies and eco-design level subsidies.

Recycling quantity subsidies are provided to recyclers based on how much material they recover. An example of such a program is the U.S. Cash for Clunkers Initiative 2009, which was expected to cost taxpayers $4 billion. Under this program, recyclers could receive subsidies between $3500 and $4500 for each end-of-life vehicle recycled [219]. On the other hand, eco-design level subsidies are provided to manufacturers based on the quantity of products they manufacture and their level of eco-design. For example, subsidies ranging from 16,000 to 29,000 yuan are offered in China for more environmentally friendly and efficient agricultural machinery [102].

A cost-based subsidy policy provides financial assistance to producers or recyclers, calculated as a percentage of their expenditures. These subsidies are categorised into eco-design cost subsidies and recycling cost subsidies. The eco-design cost subsidy supports manufacturers by covering a specified portion of their eco-design expenses, primarily incentivising eco-friendly design practices and reducing waste generation at its source. For example, in 2018, the Xiamen government in China allocated funding to green manufacturing and design initiatives, contributing less than 10% of the total investment for such projects [213]. Similarly, the governments of Australia and Canada provided subsidies covering 50% of the costs for investments in green technologies [17]. In contrast, the recycling cost subsidy focuses on reimbursing recyclers based on a percentage of their fixed costs. This approach aims to encourage increased materials recycling and improve end-of-life waste management. An example of this policy is the Shenzhen New Energy Vehicle Financial Support Policy, where the Shenzhen, China government provided recycling subsidies for battery processing, amounting to 50% of the amount verified through audits [75].

3.5.4 *Charges and Taxes*

The CE taxation framework is a comprehensive fiscal strategy that employs various policy measures aimed at different stages of a product's life cycle. These stages include (i) production, (ii) product usage, and (iii) waste management. A distinct tax mechanism is recommended for each phase to promote resource efficiency. For the production stage, a tax on virgin raw materials is proposed. This type of taxation can be implemented at three critical points within the production cycle: (i) during the extraction of raw materials, (ii) at the point of material input for initial industrial use, and (iii) at the stage of final consumption for goods that incorporate these materials [61]. While the framework does not prescribe a specific design for these interventions, policymakers are encouraged to adapt the application of these taxes to align with their respective countries' unique economic and environmental contexts.

Furthermore, fiscal and environmental considerations can support the implementation of taxes on raw material resources within public policy [56]. Economically, such taxes are often justified by either a market failure where the environmental costs of resource extraction and use are not fully accounted for by market mechanisms or the recognition that other environmental measures, such as pollution controls, may be less effective than taxation in addressing these market inefficiencies [190]. As noted by Söderholm [189], several factors influence the decision to introduce taxes on raw materials, including concerns about resource depletion, the need to address environmental externalities, efforts to mitigate future emissions or waste by targeting raw material inputs, and the intention to encourage the substitution of virgin resources with secondary or recycled materials.

The demand for raw materials often shows low price elasticity, especially in the short term [171]. This is partly because there are usually few viable alternatives to specific resources, and the processes involved in extracting and refining these materials, such as metal smelting, are highly resource-intensive. As a result, replacing primary raw materials with alternatives can be prohibitively expensive and time-consuming. Imposing a sufficiently high tax on primary raw materials could help offset the price gap between these virgin resources and secondary materials, which are generally tax-exempt. Such a measure could incentivise the use of secondary materials, provided they are readily available, thereby promoting circular production systems. However, Söderholm [189] highlighted that the effectiveness of such taxes is limited due to the low price elasticity of supply for secondary raw materials, which constrains their availability as substitutes. To enhance resource efficiency during a product's usage phase, implementing a value-added tax (VAT) reduction for reuse and repair activities has been proposed. This tax incentive aims to lower the cost of repair services, making them more accessible to consumers and encouraging the reuse of goods. By reducing the VAT rate for repairs, more local businesses could offer maintenance services, aligning with broader resource conservation goals. The EU Circular Economy Action Plan [67: 614 final] recognises the role of pricing. It states that:

> price is a key factor affecting purchasing decisions in the value chain and for final consumers. Member States are therefore encouraged to provide incentives and use economic instruments, such as taxation, to ensure that product prices better reflect environmental costs. [...] Once a product has been purchased, its lifetime can be extended through reuse and repair [...]. The reuse and repairs sectors are labour-intensive and contribute to the EU's jobs and social agenda.

Reducing VAT on repair and reuse could also mitigate environmental externalities, as prolonging product lifespans would result in significant material and energy savings [106] and reduce the need to manufacture new items [44]. Moreover, since these goods and services are not widely traded internationally, such measures are unlikely to disrupt the internal market [43]. A practical example of this policy is Sweden's reduction of VAT for certain repairs in January 2017. The VAT rate for repairing items like clothing, footwear, leather goods, and bicycles was lowered from 25 to 12% to promote reuse and repair activities [194]. Another relevant measure in Sweden is the 50% RUT tax deduction for labour costs related to home maintenance and repairs. Initially introduced in 2007 and revised in 2016, the updated legislation (Revision 2016:1055 of the income tax law 1999:1229) was implemented in 2017 alongside the reduced VAT rates for repairs [3]. Both initiatives aim to drive resource efficiency and support sustainable consumption practices.

3.6 Summary

The chapter explored various theoretical approaches to the circular economy, offering an in-depth examination of different theoretical perspectives. It also addressed circular economy policies, examining the Circular Economy Policy Framework. In addition, the discussion included key approaches to implementing circular economy policies.

References

1. Addison J (1907) The spectator. 409(65). JM Dent
2. Adeleke AQ, Bamgbade JA, Gbadebo Salimon M, Kuang Lee C (2019) Projectmanagement performance and its influence on Malaysian building projects. KnE Soc Sci. https://doi.org/10.18502/kss.v3i22.5058
3. Almén J, Dalhammar C, Milios L, Richter JL (2021) Repair in the circular economy: towards a National Swedish Strategy. In: 20th European roundtable on sustainable consumption and production. Verlag der Technischen Universität Graz, pp 21–41
4. Alvesson M, Spicer A (2019) Neo-institutional theory and organization studies: a midlife crisis? Organ Stud 40(2):199–218. https://doi.org/10.1177/0170840618772610
5. Anderson C (2005) What's the difference between policies and procedures? https://www.bizmanualz.com/writing-policies-and-procedures/whats-the-differencebetween-policies-and-procedures.html
6. Aoi C, de Coning CH, Thakur R (2007) The unintended consequences of peacekeeping operations. United Nations University Press, Tokyo
7. Appolloni A, D'Adamo I, Gastaldi M, Santibanez-Gonzalez EDR, SettembreBlundo D (2021) Growing e-waste management risk awareness points towards new recycling scenarios: the view of the Big Four's youngest consultants. Environ Technol Innov 23:101716. https://doi.org/10.1016/j.eti.2021.101716
8. Artiach T, Lee D, Nelson D, Walker J (2010) The determinants of corporate sustainability performance. Account Fin 50(1):31–51. https://doi.org/10.1111/j.1467-629X.2009.00315.x
9. Asari M, Sakai SI (2013) Li-ion battery recycling and cobalt flow analysis in Japan. Resour Conserv Recycl 81:52–59
10. Ayres RU, Kneese AV (1969) Production, consumption, and externalities. Am Econ Rev 59(3):282–297
11. Azevedo SG, Godina R, Matias JCDO (2017) Proposal of a sustainable circular index for manufacturing companies. Resources 6(4):63
12. Bailey I (2002) European environmental taxes and charges: economic theory and policy practice. Appl Geogr 22(3):235–251. https://doi.org/10.1016/S0143-6228(02)00011-5
13. Barbosa LC (2009) The rise of brazil as an agricultural superpower and the dilemma over the amazon rain forest: a treadmill of production view. SSRN Electron J. https://doi.org/10.2139/ssrn.1338937
14. Baumol WJ (1970) A new rationale for corporate social policy. Heath Lexington Books. https://cir.nii.ac.jp/crid/1130000795378231040.bib?lang=en
15. Bedau MA (1997) Weak emergence. Philos Perspect 11:375–399
16. Bedau MA, Humphreys PE (2008) Emergence: contemporary readings in philosophy and science. MIT Press, Boston
17. Bian J, Zhang G, Zhou G (2020) Manufacturer vs. consumer subsidy with green technology investment and environmental concern. Eur J Oper Res 287(3):832–843

18. Bocken NMP, Short SW, Rana P, Evans S (2014) A literature and practice review to develop sustainable business model archetypes. J Clean Prod 65:42–56. https://doi.org/10.1016/j.jclepro.2013.11.039
19. Bocken NMP, de Pauw I, Bakker C, Van der Grinten B (2016) Product design and business model strategies for a circular economy. J Indus Prod Eng 33(5):308–320. https://doi.org/10.1080/21681015.2016.1172124
20. Bolger K, Doyon A (2019) Circular cities: exploring local government strategies to facilitate a circular economy. Eur Plan Stud 27(11):2184–2205. https://doi.org/10.1080/09654313.2019.1642854
21. Bonciu F (2014) The European economy: from a linear to a circular economy. Romanian J Eur Aff 14:78
22. Borel-Saladin JM, Turok IN (2013) The green economy: incremental change or transformation? Environ Policy Gov 23(4):209–220. https://doi.org/10.1002/eet.1614
23. Boulding KE (1966) The economics of coming spaceship earth. In: Jarrett H (ed) Environmental quality in a growing economy. John Hopkins, pp 3–14
24. Bowen H (1953) Social responsibilities of the businessman. Harper, New York
25. Braun AT, Kleine-Möllhoff P, Reichenberger V, Seiter S (2018) Survey concerning enablers for material efficiency activities in manufacturing, their supply chains and the transformation towards circular economy. https://www.econstor.eu/handle/10419/178639
26. Burchell J, Cook J (2013) Sleeping with the enemy? strategic transformations in business–NGO relationships through stakeholder dialogue. J Bus Ethics 113(3):505–518. https://doi.org/10.1007/s10551-012-1319-1
27. Camacho-Otero J, Boks C, Pettersen IN (2018) Consumption in the circular economy: a literature review. Sustainability 10(8):2758
28. Campiglio L, Pineschi L, Siniscalco D, Treves T (1994) The environment after Rio: international law and economics
29. Canadian Council of Ministers of the Environment (CCME) (2009) Canada-wide action plan for extended producer responsibility. https://www.Ccme.ca/Files/Resources/Waste/Extended/Pn_1499_epr_cap_e.Pdf
30. Carroll AB (1999) Corporate social responsibility. Bus Soc 38(3):268–295. https://doi.org/10.1177/000765039903800303
31. Castka P, Prajogo D (2013) The effect of pressure from secondary stakeholders on the internalization of ISO 14001. J Clean Prod 47:245–252. https://doi.org/10.1016/j.jclepro.2012.12.034
32. CE & BioIS (2014) Study on modelling of the economic and environmental impacts of raw material consumption
33. Chan S-C, Fang W (2007) Antecedents and performance outcomes of environmental management. In: International conference on business and information. Tokyo, Japan
34. Chen M (2005) End-of-life vehicle recycling in China now and the future. http://www.springerlink.vom/Content/58543mn7682747g7/Fulltext.Pdf?Page=1(En)
35. Cherry CE, Pidgeon NF (2018) Is sharing the solution? Exploring public acceptability of the sharing economy. J Clean Prod 195:939–948. https://doi.org/10.1016/j.jclepro.2018.05.278
36. Chi NK (2009) 3R policy development and solid waste management status in Vietnam. In: International workshop on 3R and waste management
37. Choate A (2008) Landfill analysis overview. EPA Organics Workshop. http://epa.Gov/Climatechange/Wycd/Waste/Downloads/Landfill_Briefing_final.Pdf
38. Collect-a-Can (2011) A walk through history with collect-a-can. http://collectacan.co.za/index.php?option=com_content&view=article&id=128:march2011-a-walk-through-history-with-collect-a-can&catid=5:junepres
39. Collect-a-Can (2015) Beverage can recovery rate 1993 to 2015. http://www.collectacan.co.za/index.php/can-recycling/recovery-rate
40. Colquitt JA, Zapata-Phelan CP (2007) Trends in theory building and theory testing: a five-decade study of the academy of management journal. Acad Manag J 50(6):1281–1303. https://doi.org/10.5465/amj.2007.28165855

41. Commoner B (1971) The closing circle: nature, man, technology. Knopf, Alfred A
42. ComplianceBridge (2017) The key difference between policy vs. procedures. https://www.bizmanualz.com/writing-policies-and-procedures/whats-the-differencebetween-policies-and-procedures.html
43. Copenhagen Economics (2007) Study on reduced VAT applied to goods and services in the member states of the European Union. https://taxationcustoms.ec.europa.eu/system/files/2016-09/study_reduced_vat.pdf
44. Cooper DR, Gutowski TG (2017) The environmental impacts of reuse: a review. J Ind Ecol 21(1):38–56
45. Cormier D, Magnan M (2017) Does a firm's exposure to ethical failures matter to financial markets? A governance perspective. J Manage Governance 21(2):267–289. https://doi.org/10.1007/s10997-016-9341-9
46. Crutzen PJ (2016) Geology of mankind, pp 211–215. https://doi.org/10.1007/978-3-319-27460-7_10
47. Daddi T, Testa F, Frey M, Iraldo F (2016) Exploring the link between institutional pressures and environmental management systems effectiveness: an empirical study. J Environ Manage 183:647–656. https://doi.org/10.1016/j.jenvman.2016.09.025
48. Dai R, Zhang J, Tang W (2017) Cartelization or cost-sharing? Comparison of cooperation modes in a green supply chain. J Clean Prod 156:159–173. https://doi.org/10.1016/j.jclepro.2017.04.011
49. Darby L, Obara L (2005) Household recycling behaviour and attitudes towards the disposal of small electrical and electronic equipment. Resour Conserv Recycl 44(1):17–35. https://doi.org/10.1016/j.resconrec.2004.09.002
50. Darnall N, Henriques I, Sadorsky P (2008) Do environmental management systems improve business performance in an international setting? J Int Manag 14(4):364–376. https://doi.org/10.1016/j.intman.2007.09.006
51. Darwin C (1859) On the origins of species by means of natural selection. Murray
52. Das TK, Teng B (2002) The dynamics of alliance conditions in the alliance development process. J Manage Stud 39(5):725–746. https://doi.org/10.1111/1467-6486.00006X
53. DiMaggio P (1988) Interest and agency in institutional theory. In: Zucker L (ed) Institutional patterns and culture. Ballinger Publishing Company, Cambridge, pp 3–22
54. DiMaggio PJ, Powell WW (1983) The iron cage revisited: institutional isomorphismand collective rationality in organizational fields. Am Sociol Rev 48(2):147. https://doi.org/10.2307/2095101
55. Dobbs R, Oppenheim J, Thompson F, Brinkman M, Zornes M (2011) Resource revolution: meeting the world's energy. materials, food, and water needs. https://www.mckinsey.com/capabilities/sustainability/our-insights/resource-revolution
56. Domenech T, Bahn-Walkowiak B (2019) Transition towards a resource efficient circular economy in Europe: policy lessons from the EU and the member states. Ecol Econ 155:7–19
57. Douvan AR (2009). Italian approaches on 3R developments. In: International workshop on 3R and waste management. Kyoto City
58. Dresner S (2012) The principles of sustainability. Routledge. https://doi.org/10.4324/9781849773249
59. Driedger RJ (2001) From cradle to grave: extended producer responsibility for household hazardous wastes in british columbia. J Ind Ecol 5(2):89–102. https://doi.org/10.1162/10881980152830150
60. Dubey R, Gunasekaran A, Childe SJ, Papadopoulos T, Helo P (2019) Supplierrelationship management for circular economy. Manag Decis 57(4):767–790. https://doi.org/10.1108/MD-04-2018-0396
61. Eckermann F, Golde M, Herczeg M, Mazzanti M, Zoboli R, Speck S (2016) Material resource taxation. https://www.eionet.europa.eu/etcs/etc-wmge/products/etc-wmgereports/material-resource-taxation-2013-an-analysis-for-selected-material-resources
62. Elkington J (1997) Cannibals with forks: the triple bottom line of 21st century business. Oxford

63. Ellen MacArthur Foundation (2013) Towards the circular economy. file: ///C:/Users/USER/Downloads/Towards_A_Circular_Economy.pdf. Accessed 03 March, 2024
64. Ellen MacArthur Foundation (2015) Growth within: a circular economy vision for a competitive Europe. Ellen MacArthur Foundation 100:81–92
65. Elliott SR (2005) Sustainability: an economic perspective. Resour, Conserv Recycl 44(3):263–277
66. European Commission (2010) A European strategy for smart, sustainable and inclusive growth. https://eurlex.europa.eu/lexuriserv/lexuriserv.do?uri=com:2010:2020:fin:en:pdf
67. European Commission (2015) An EU action plan for the circular economy. https://eur-lex.europa.eu/resource.html?uri=cellar:8a8ef5e8-99a0-11e5-b3b701aa75ed71a1.0012.02/doc_1&format=pdf
68. European Union (2008) Directive 2008/98/EC of the European Parliament and the Council of 19 November 2008 on Waste and Repealing Certain Directives. Official J Eur Union, 22/11/2008
69. European Council Directive (1999) 1999/31/EC of 26 April 1999 on the landfill of waste. Official J L 282:0016
70. European Environment Agency (2009) Diverting waste from landfill—effectiveness of waste management policies in the European Union. http://www.eea.europa.eu/publications/diverting-waste-from-landfill-effectiveness-ofwaste-management-policies-in-the-european-union/at_download/file (en)
71. Farnsworth KD, Niklas KJ (1995) Theories of optimization, form and function in branching architecture in plants. Funct Ecol 9(3):355–363
72. Favot M, Massarutto A (2019) Rare-earth elements in the circular economy: the case of yttrium. J Environ Manage 240:504–510. https://doi.org/10.1016/j.jenvman.2019.04.002
73. Ferrão PC (2002) Economy's metabolism: Indicators, scales, and technology. In: Proceedings of the 6th international conference on technology policy and innovation
74. Ferrão P, Ribeiro P, Silva P (2008) A management system for end-of-life tyres: A Portuguese case study. Waste Manage 28(3):604–614. https://doi.org/10.1016/j.wasman.2007.02.033
75. Finance Commission of Shenzhen Municipality (2018) Development and reform commission of Shenzhen municipality (DRCSM) financial support policy for promotion and application of new energy vehicles in Shenzhen in 2018. http://Www.Sz.Gov.Cn/Zfgb/2019/Gb1087/Content/Mpost_4998331.Html
76. Fischer C (2009) EU waste policy: development and achievements. In: International workshop on 3R and waste management, Kyoto City
77. Folke C, Carpenter SR, Walker B, Scheffer M, Chapin T, Rockström J (2010) Resilience thinking: integrating resilience, adaptability and transformability. Ecol Soc 15(4):1–9
78. Forslind KH (2005) Implementing extended producer responsibility: the case of Sweden's car scrapping scheme. J Clean Prod 13(6):619–629. https://doi.org/10.1016/j.jclepro.2003.12.017
79. Freeman RE (1984) Strategic management. Cambridge University Press. https://doi.org/10.1017/CBO9781139192675
80. Freeman RE (1984b) Strategic management: a stakeholder approach. Pitman
81. Friedman M (1962) Capitalism and freedom. University of Chicago Press
82. Friedman M (1970) The social responsibility of business is to increase its profits. New Y Mag 13:32–33
83. Galvão GDA, De Nadae J, Clemente DH, Chinen G, De Carvalho MM (2018) Circular economy: overview of barriers. Procedia Cirp 73:79–85
84. Geels FW (2012) A socio-technical analysis of low-carbon transitions: introducing the multi-level perspective into transport studies. J Transp Geogr 24:471–482. https://doi.org/10.1016/j.jtrangeo.2012.01.021
85. Geng Y, Doberstein B (2008) Developing the circular economy in China: challenges and opportunities for achieving "leapfrog development." Int J Sust Dev World 15(3):231–239
86. Gheitarani F, Nawaser K, Hanifah H, Vafaei-Zadeh A (2023) Human-behavioural micro-foundations of dynamic capabilities: a systematic review of the last two decades of research. Int J Manag Decis Mak 22(1):1–26

87. Gioia DA, Pitre E (1990) Multiparadigm perspectives on theory building. Acad Manag Rev 15(4):584–602
88. Global Reporting Initiative (2002) The GRI Standards: a guide for policy makers. https://www.globalreporting.org/media/nmmnwfsm/gri-policymakers-guide.pdf
89. Gong H, Wang MQ, Wang H (2013) New energy vehicles in China: policies, demonstration, and progress. Mitig Adapt Strat Glob Change 18:207–228
90. Gordon RB, Bertram M, Graedel TE (2006) Metal stocks and sustainability. Proc Natl Acad Sci 103(5):1209–1214. https://doi.org/10.1073/pnas.0509498103
91. Gorji M-A, Jamali M-B, Iranpoor M (2021) A game-theoretic approach for decision analysis in end-of-life vehicle reverse supply chain regarding government subsidy. Waste Manage 120:734–747. https://doi.org/10.1016/j.wasman.2020.10.043
92. Gould KA, Lewis TL (2009) Twenty lessons in environmental sociology. Oxford University Press
93. Green Brain (2022) Green Brain empowers growers to make better irrigation decisions. https://greenbrain.ag/
94. Gutowski T, Murphy C, Allen D, Bauer D, Bras B, Piwonka T, Wolff E et al (2005) Environmentally benign manufacturing: observations from Japan, Europe and the United States. J Clean Prod 13(1):1–17
95. Hartley K, Van Santen R, Kirchherr J (2020) Policies for transitioning towards a circular economy: expectations from the European Union (EU). Resour Conserv Recycl 155:104634. https://doi.org/10.1016/j.resconrec.2019.104634
96. Hashemzadeh GR, Khaksar SMS, Nawaser K, Jahanshahi AA (2011) Technological dimension of customer relationship management. Indian J Sci Technol 4(11):1565–1572
97. Hasnas J (1998) The normative theories of business ethics: a guide for the perplexed. Bus Ethics Q 8(1):19–42. https://doi.org/10.2307/3857520
98. Hathaway R (2009) UK policy developments in support of 3R. In: International workshop on 3R and waste management, Kyoto City
99. Heinrichs H (2013) Sharing economy: a potential new pathway to susainability. GAIA—Ecol Persp Sci Soc 22(4):228–231. https://doi.org/10.14512/gaia.22.4.5
100. Hollins O (2017) Business resource efficiency quantification of the no cost/low cost resource efficiency opportunities in the UK economy in 2014, s.l.: Department for Environment, Food and Rural Affairs. Retrieved from http://sciencesearch.defra.gov.uk/Document.aspx?Document=14112_BusinessResourceEfficiency.pdf
101. Hopwood B, Mellor M, O'Brien G (2005) Sustainable development: mapping different approaches. Sustain Dev 13(1):38–52. https://doi.org/10.1002/sd.244
102. Huang W, Zhou W, Chen J, Chen X (2019) The government's optimal subsidy scheme under manufacturers' competition of price and product energy efficiency. Omega 84:70–101
103. Huang Y-C, Yang M-L (2014) Reverse logistics innovation, institutional pressures and performance. Manag Res Rev 37(7):615–641. https://doi.org/10.1108/MRR-03-2013-0069
104. Hung Lau K, Wang Y (2009) Reverse logistics in the electronic industry of China: a case study. Supply Chain Manage: Int J 14(6):447–465
105. International Energy Agency (IEA) (2012) World energy outlook. https://www.iea.org/reports/world-energy-outlook-2012
106. International Resource Panel (2018) Re-defining value—the manufacturing revolution. Remanufacturing, refurbishment. In: Repair and direct reuse in the circular economy. United Nations Environment Programme. https://www.resourcepanel.org/reports/redefining-value-manufacturing-revolution Accessed 4 February, 2024
107. Islam MS (2013) Development, power, and the environment: neoliberal paradox in the age of vulnerability. Routledge
108. Islam S, Hossain I (2015) The global treadmill of production and the environment. In: Social justice in the globalization of production. Palgrave Macmillan UK, pp 144–158. https://doi.org/10.1057/9781137434012_9
109. Jacobs M (1997) Sustainability and markets: on the neoclassical model of environmental economics. New Political Economy 2(3):365–385. https://doi.org/10.1080/13563469708406313

110. Jamali D (2006) Insights into triple bottom line integration from a learning organization perspective. Bus Process Manag J 12(6):809–821. https://doi.org/10.1108/14637150610710945
111. Jänicke M (2012) "Green growth": from a growing eco-industry to economic sustainability. Energy Policy 48:13–21. https://doi.org/10.1016/j.enpol.2012.04.045
112. Japan Ministry of the Environment (2008) Annual report on the environment and the sound material-cycle society in Japan 2008. https://www.env.go.jp/en/wpaper/2008/index.html
113. Jennings PD, Zandbergen PA (1995) Ecologically sustainable organizations: an institutional approach. Acad Manag Rev 20(4):1015–1052
114. Jorgenson AK (2009) The sociology of unequal exchange in ecological context: a panel study of lower-income countries, 1975–2000 1. Sociol Forum 24(1):22–46. https://doi.org/10.1111/j.1573-7861.2008.01085.x
115. Jia F, Zuluaga-Cardona L, Bailey A, Rueda X (2018) Sustainable supply chain management in developing countries: an analysis of the literature. J Clean Prod 189:263–278. https://doi.org/10.1016/j.jclepro.2018.03.248
116. Kauppi K (2013) Extending the use of institutional theory in operations and supply chain management research. Int J Oper Prod Manag 33(10):1318–1345. https://doi.org/10.1108/IJOPM-10-2011-0364
117. Kauppi K, Hannibal C (2017) Institutional pressures and sustainability assessment in supply chains. Supply Chain Manage: Int J 22(5):458–472. https://doi.org/10.1108/SCM-01-2017-0004
118. Kezar A, Bernstein-Sierra S (2019) Examining processes of normative isomorphism and influence in scaled change among higher education intermediary organizations. AERAOpen 5(4):233285841988490. https://doi.org/10.1177/2332858419884905
119. Kibert NC (2004) Extended producer responsibility: a tool for achieving sustainable development. J Land Use Environ Law 19(2):503–523
120. Korhonen J, Nuur C, Feldmann A, Birkie SE (2018) Circular economy as an essentially contested concept. J Clean Prod 175:544–552. https://doi.org/10.1016/j.jclepro.2017.12.111
121. Koza MP, Lewin AY (1998) The co-evolution of strategic alliances. Organ Sci 9(3):255–264. https://doi.org/10.1287/orsc.9.3.255
122. Labuschagne C, Brent AC, Van Erck RP (2005) Assessing the sustainability performances of industries. J Clean Prod 13(4):373–385
123. Langrova V (2002) Comparative analysis of EPR programmes for small consumer batteries: case study of the Netherlands, Switzerland and Sweden. IIIEE, Lund University. http://www.iiiee.lu.se/Publication.nsf/$webAll/F8439420F8F74B7EC1256CCC002E3532/$FILE/Report_2002_9_rev.pdf
124. Lara VHM (2016) Exploring EU and non-state actors negotiation dynamics in FTAS: the case of the stakeholders' events and the energy sector in the transatlantic trade and investment partnership (TTIP). Unpublished masters dissertation
125. Law for the Promotion of Effective Utilization of Resources (2001). https://www.env.go.jp/content/900452886.pdf
126. Lee MP (2008) A review of the theories of corporate social responsibility: its evolutionary path and the road ahead. Int J Manag Rev 10(1):53–73. https://doi.org/10.1111/j.1468-2370.2007.00226.x
127. Leendertse PW (2016) The transition from a linear to circular economy: an innovation system analysis of the composites industry. https://studenttheses.uu.nl/handle/20.500.12932/22343
128. Lenz T (2017) Alexandru Grigorescu. 2015. Democratic intergovernmental organizations? Normative pressures and decision-making rules. Cambridge University Press, New York, NY
129. Lewandowski M (2016) Designing the business models for circular economy—towards the conceptual framework. Sustainability 8(1):43. https://doi.org/10.3390/su8010043
130. Lewin AY, Long CP, Carroll TN (1999) The coevolution of new organizational forms. Organ Sci 10(5):535–550. https://doi.org/10.1287/orsc.10.5.535
131. Li J (2009) Summary policy and legislative tools in promoting waste management under circular economic model in China. In: International workshop on 3R and waste management. Kyoto City

132. Li Y (2014) Environmental innovation practices and performance: moderating effect of resource commitment. J Clean Prod 66:450–458
133. Lindhqvist T (2000) Extended producer responsibility in cleaner production: policy principle to promote environmental improvements of product systems. The International Institute for Industrial Environmental Economics. IIIEE, Lund University. https://lup.lub.lu.se/search/files/4433708/1002025.pdf
134. Linnenluecke MK, Griffiths A (2013) Firms and sustainability: mapping the intellectual origins and structure of the corporate sustainability field. Glob Environ Chang 23(1):382–391. https://doi.org/10.1016/j.gloenvcha.2012.07.007
135. Liu Y, Quan B, Xu Q, Forrest JY-L (2019) Corporate social responsibility and decision analysis in a supply chain through government subsidy. J Clean Prod 208:436–447. https://doi.org/10.1016/j.jclepro.2018.10.121
136. Liuyang Management Committee of Manufacturing Industry (2010) Several provisions of supporting remanufacturing enterprises development and preferential policies. http://Shop.Ebdoor.Com/Shops/750369/DynamicInfos/999111.Aspx
137. Loorbach D, Wijsman K (2013) Business transition management: Exploring a new role for business in sustainability transitions. J Clean Prod 45:20–28. https://doi.org/10.1016/j.jclepro.2012.11.002
138. López-Gamero MD, Molina-Azorín JF, Claver-Cortés E (2010) The potential of environmental regulation to change managerial perception, environmental management, competitiveness and financial performance. J Clean Prod 18(10–11):963–974. https://doi.org/10.1016/j.jclepro.2010.02.015
139. Lozano R, Carpenter A, Huisingh D (2015) A review of 'theories of the firm' and their contributions to corporate sustainability. J Clean Prod 106:430–442. https://doi.org/10.1016/j.jclepro.2014.05.007
140. Mangla SK, Luthra S, Mishra N, Singh A, Rana NP, Dora M, Dwivedi Y (2018) Barriers to effective circular supply chain management in a developing country context. Prod Plann Control 29(6):551–569. https://doi.org/10.1080/09537287.2018.1449265
141. Manninen K, Koskela S, Antikainen R, Bocken N, Dahlbo H, Aminoff A (2018) Do circular economy business models capture intended environmental value propositions? J Clean Prod 171:413–422. https://doi.org/10.1016/j.jclepro.2017.10.003
142. Markley MJ, Davis L (2007) Exploring future competitive advantage through sustainable supply chains. Int J Phys Distrib Logist Manag 37(9):763–774. https://doi.org/10.1108/09600030710840859
143. Marx K (1981) Capital, vol 3. Penguin
144. Masi D, Kumar V, Garza-Reyes JA, Godsell J (2018) Towards a more circular economy: exploring the awareness, practices, and barriers from a focal firm perspective. Prod Plann Control 29(6):539–550. https://doi.org/10.1080/09537287.2018.1449246
145. Mathiyazhagan K, Govindan K, NoorulHaq A, Geng Y (2013) An ISM approach for the barrier analysis in implementing green supply chain management. J Clean Prod 47:283–297. https://doi.org/10.1016/j.jclepro.2012.10.042
146. McKerlie K, Knight N, Thorpe B (2006) Advancing extended producer responsibility in Canada. J Clean Prod 14(6–7):616–628. https://doi.org/10.1016/j.jclepro.2005.08.001
147. Mebratu D (1998) Sustainability and sustainable development. Environ Impact Assess Rev 18(6):493–520. https://doi.org/10.1016/S0195-9255(98)00019-5
148. Meyer J, Rowan B (1977) Institutionalized organizations: Formal structures as myth and ceremony. Am J Sociol 83(1):340–363
149. Ministry of Environment, Republic of Korea (2008) The national consciousness census for environmental conservation. http://webbook.me.go.kr/DLi-File/pdf/2009/11/183579.pdf
150. Missimer M, Robèrt K-H, Broman G (2017) A strategic approach to social sustainability—Part 1: exploring the social system. J Clean Prod 140:32–41. https://doi.org/10.1016/j.jclepro.2016.03.170
151. Missimer M, Robèrt K-H, Broman G (2017) A strategic approach to social sustainability—Part 2: a principle based definition. J Clean Prod 140:42–52. https://doi.org/10.1016/j.jclepro.2016.04.059

152. Naess P (2001) Urban planning and sustainable development. Eur Plan Stud 9(4):503–524
153. Nahman A (2010) Extended producer responsibility for packaging waste in South Africa: current approaches and lessons learned. Resour Conserv Recycl 54(3):155–162. https://doi.org/10.1016/j.resconrec.2009.07.006
154. National Aeronautics and Space Administration (2022) Global climate change. https://climate.nasa.gov/vital-signs/global-temperature/
155. Nicolis G, Prigogine I, Freeman WH (1989) Exploring complexity. An introduction. W H Freeman & Company
156. Nnorom IC, Osibanjo O (2008) Overview of electronic waste (e-waste) management practices and legislations, and their poor applications in the developing countries. Resour Conserv Recycl 52(6):843–858. https://doi.org/10.1016/j.resconrec.2008.01.004
157. Organisation for Economic Co-operation & Development (OECD) (2011) Towards green growth: monitoring progress: OECD indicators. OECD Publishing, Paris
158. OECD (2017) Employment implications of green growth: linking jobs, growth, and green policies. https://www.oecd.org/environment/employment-implications-of-greengrowth-oecd-report-g7-environment-ministers.pdf
159. OECD (2001) Extended producer responsibility: a guidance manual for governments. https://www.oecd-ilibrary.org/environment/extended-roducerresponsibility_9789264189867-en
160. Pan S-Y, Gao M, Kim H, Shah KJ, Pei S-L, Chiang P-C (2018) Advances and challenges in sustainable tourism toward a green economy. Sci Total Environ 635:452–469. https://doi.org/10.1016/j.scitotenv.2018.04.134
161. Park JW (2009) 3R policies of Korea. Ministry of Environment. https://eng.me.go.kr/board.do?method=view&docseq=195&bbscode=res_mat_policy (en)
162. Pearce D (1992) Green economics. Environ Values 1(1):3–13
163. Pearce D, Atkinson G (1992) Are national economies sustainable. Measuring SD
164. Pearce DW, Turner RK (1989) Economics of natural resources and the environment. Johns Hopkins University Press
165. Perry II JF (2012). The impact of supply chain business processes on competitive advantageand organizational performance. https://apps.dtic.mil/sti/citations/ADA559397
166. Pezzey JC (1992) Economic analysis of sustainable growth and sustainable development. https://documents.worldbank.org/en/publication/documents-570
167. Pociovălişteanu D, Novo-Corti I, Aceleanu M, Şerban A, Grecu E (2015) Employment policies for a green economy at the european union level. Sustainability 7(7):9231–9250. https://doi.org/10.3390/su7079231
168. Pörtner HO (2021) Climate impacts on organisms, ecosystems and human societies: integrating OCLTT into a wider context. J Exp Biol 224(Suppl_1):jeb238360
169. Prieto-Sandoval V, Jaca C, Ormazabal M (2018) Towards a consensus on the circular economy. J Clean Prod 179:605–615
170. Qiao H, Su Q (2021) Impact of government subsidy on the remanufacturing industry. Waste Manage 120:433–447. https://doi.org/10.1016/j.wasman.2020.10.005
171. Radetzki M, Aguilera RF (2008) A handbook of primary commodities in the global economy. Energy J 29(4):168
172. Ravetz J (2016) City-region 2020: integrated planning for a sustainable environment. Routledge
173. Redclift M (2005) Sustainable development (1987–2005): an oxymoron comes of age. Sustain Dev 13(4):212–227. https://doi.org/10.1002/sd.281
174. Reid D (2005) Sustainable development: an introductory guide. Earthscan Publications
175. Republic of South Africa (2012) Integrated industry waste tyre management plan of the recycling and economic development initiative of South Africa (REDISA). Government Gazette No. 35927, Notice No. 988. 2012. http://sawic.environment.gov.za/documents/1702.pdf
176. Rodrigues S, Child J (2003) Co-evolution in an institutionalized environment. J Manage Stud 40(8):2137–2162. https://doi.org/10.1046/j.1467-6486.2003.00415.x
177. Rosa P, Sassanelli C, Terzi S (2019) Circular business models versus circular benefits: an assessment in the waste from electrical and electronic equipments sector. J Clean Prod 231:940–952. https://doi.org/10.1016/j.jclepro.2019.05.310

178. Roxas B, Coetzer A (2012) Institutional environment, managerial attitudes and environmental sustainability orientation of small firms. J Bus Ethics 111(4):461–476. https://doi.org/10.1007/s10551-012-1211-z
179. Sacirovic S, Ketin S, Vignjevic N (2019) Eco-industrial zones in the context of sustainability development of urban areas. Environ Sci Pollut Res 26(24):24346–24356. https://doi.org/10.1007/s11356-018-1390-y
180. Saeed A, Jun Y, Nubuor S, Priyankara H, Jayasuriya M (2018) Institutional pressures, green supply chain management practices on environmental and economic performance: a two theory view. Sustainability 10(5):1517. https://doi.org/10.3390/su10051517
181. Salmenperä H, Pitkänen K, Kautto P, Saikku L (2021) Critical factors for enhancing the circular economy in waste management. J Clean Prod 280:124339. https://doi.org/10.1016/j.jclepro.2020.124339
182. Schnaiberg A (1980) The environment: from surplus to scarcity. Oxford University Press
183. Schröder P, Lemille A, Desmond P (2020) Making the circular economy work for human development. Resour Conserv Recycl 156:104686. https://doi.org/10.1016/j.resconrec.2020.104686
184. Scott WR (1987) Institutional theory. Adm Sci Q 32:493–511
185. Scupola A (2003) The adoption of internet commerce by smes in the south of Italy: an environmental, technological and organizational perspective. J Glob Inf Technol Manag 6(1):52–71. https://doi.org/10.1080/1097198X.2003.10856343
186. Sino-Japan Friendship Centre for Environmental Protection (2008) Base for environmental cooperation for China. https://www.mofa.go.jp/policy/oda/category/environment/pamph/2001/coop-1.html
187. Slaper TF, Hall TJ (2011) The triple bottom line: what is it and how does it work. Indiana Bus Rev 86(1):4–8
188. Smith A (1776) The wealth of nations. Econ Inquiry 15(3), 309–325
189. Söderholm P (2011) Taxing virgin natural resources: lessons from aggregates taxation in Europe. Resour Conserv Recycl 55(11):911–922
190. Söderholm P, Tilton JE (2012) Material efficiency: an economic perspective. Resour Conserv Recycl 61:75–82
191. South African Waste Information Centre (SAWIC) (2014) Approach to waste in South Africa. https://sawic.environment.gov.za/?Menu=60
192. Steurer R, Langer ME, Konrad A, Martinuzzi A (2005) Corporations, stakeholders and sustainable development I: a theoretical exploration of business–society relations. J Bus Ethics 61(3):263–281. https://doi.org/10.1007/s10551-005-7054-0
193. Sutton RI, Staw BM (1995) What theory is not. Adm Sci Q 40(3):371–384
194. Sweden Government (2017) From value chain to value cycle—this is how Sweden gets a more circular economy. https://Www-Regeringen-Se.Translate.Goog/Rattsliga-Dokument/Statens
195. Tsoi J (2010) Stakeholders' perceptions and future scenarios to improve corporate social responsibility in Hong Kong and Mainland China. J Bus Ethics 91:391–404
196. Turok I (2012) Urbanisation and development in South Africa: economic imperatives, spatial distortions and strategic responses. https://www.iied.org/sites/default/files/pdfs/migrate/10621IIED.pdf
197. United Nations (2013) Green growth and green economy. http://www.unescap.org/ourwork/environment-development/green-growth-green-economy/about2013
198. UNEP (2015) Guidance on organizational life cycle assessment. Life-cycle initiative. United Nations Environment Programme and Society for Environmental Toxicology and Chemistry, Paris, France. http://www.lifecycleinitiative.org/wp-content/uploads/2015/04/o-lca_24.4.15-web.pdf
199. Unterseh TS (2009) Extended producer responsibility—The German experience. In: International workshop on 3R and waste management, Kyoto City, 2009
200. U.S. Environmental Protection Agency (2016) Deconstruction and materials reuse: a guide for building professionals. https://delta-nstitute.org/wpcontent/uploads/2018/05/Deconstruction-Go-Guide-6-13-18-.pdf

201. Van Buren N, Demmers M, Van der Heijden R, Witlox F (2016) Towards a circular economy: the role of Dutch logistics industries and governments. Sustainability 8(7):647
202. Van Ewijk S (2018) Resource efficiency and the circular economy: concepts, economic benefits, barriers, and policies. https://discovery.ucl.ac.uk/id/eprint/10054117/
203. van Marrewijk M (2003) Multiple levels of corporate sustainability. J Bus Ethics 44(2):107–119. https://doi.org/10.1023/A:1023383229086
204. Victor P, Hanna S, Kubursi A (1998) How strong is weak sustainability? Pp 195–210. https://doi.org/10.1007/978-94-017-3188-1_12
205. Vietnam Environment Administration (2023) Vietnam environment administration. http://vea.gov.vn/vn/pages/trangchu.aspx (vn)
206. Vietnam Government (2009) National strategy on integrated SWM to 2025 and vision to 2050 Vietnam (2149/QD-TTg). http://vea.gov.vn/VN/vanbanphapquy/Pages/ChiếnlượcQuốcgiavềquảnlýtổnghợpchấtthảirắnđếnnăm2025,tầmnhìnđếnnăm2050.aspx (Vn)
207. Wallerstein I (1974) The modern world-system i capitalist agriculture and the origins of the european world-economy in the sixteenth century, with a new prologue. University of California Press
208. Walls M (2006) EPR policies and product design: Economic theory and selected case studies. RFF Discussion Paper No. 06-08. https://doi.org/10.2139/ssrn.901661
209. Walz R (2011) Employment and structural impacts of material efficiency strategies: results from five case studies. J Clean Prod 19(8):805–815. https://doi.org/10.1016/j.jclepro.2010.06.023
210. Wang Z, Obata N, Zhou W (2007) The current status and prob- lem of automobile recycling project in China on comparison of a China-Japan. Seisaku Kagaku 15(1):83–97
211. WCED (1987) Our common future. World Comm Environ Dev 17(1):1–91
212. Widmer R, Oswald-Krapf H, Sinha-Khetriwal D, Schnellmann M, Böni H (2005) Global perspectives on e-waste. Environ Impact Assess Rev 25(5):436–458. https://doi.org/10.1016/j.eiar.2005.04.001
213. Xiamen Government (2018) Measures for the administration of municipal-level special funds for energy conservation and circular economy development in Xiamen. http://www.xm.gov.cn/zfxxgk/xxgkznml/szfgz/bmgfwj/201808/t20180828_2108086.htm
214. Xue X, Shen Q, Ren Z (2010) Critical review of collabourative working in construction projects: business environment and human behaviours. J Manag Eng 26(4):196–208. https://doi.org/10.1061/(ASCE)ME.1943-5479.0000025
215. Yuan Z, Bi J, Moriguichi Y (2006) The circular economy: a new development strategy in China. J Ind Ecol 10(1–2):4–8. https://doi.org/10.1162/108819806775545321
216. Zhijun F, Nailing Y (2007) Putting a circular economy into practice in China. Sustain Sci 2(1):95–101. https://doi.org/10.1007/s11625-006-0018-1
217. Zhu Q, Sarkis J (2007) The moderating effects of institutional pressures on emergent green supply chain practices and performance. Int J Prod Res 45(18–19):4333–4355. https://doi.org/10.1080/00207540701440345
218. Zhu Q, Sarkis J, Lai K (2008) Green supply chain management implications for "closing the loop." Transp Res Part E: Logist Transp Rev 44(1):1–18. https://doi.org/10.1016/j.tre.2006.06.003
219. Zolnik EJ (2012) Estimates of statewide and nationwide carbon dioxide emission reductions and their costs from cash for clunkers. J Transp Geogr 24:271–281

Chapter 4
The Theoretical Frameworks of the Circular Economy Business Model (CEBM)

Abstract An analysis of existing literature highlights that while organisations within a particular industry may adopt similar components of CEBMs, these models are uniquely tailored to individual firms rather than applying uniformly across an entire industry. Furthermore, CEBMs are characterised by their market-specific focus and entrepreneurial ventures. They are inherently dynamic, adapting over time to align with changes in the business environment. The literature also challenges the assumption that adopting CE practices in the construction sector is complicated due to its fragmented nature. Instead, it reveals that CEBMs within the construction industry are similarly segmented in alignment with the CE framework. This segmentation presents opportunities for entrepreneurs in various sub-sectors of the construction industry to innovate, deliver value, and meet client needs effectively. Furthermore, the review highlights that the CEBMs examined were developed using the Business Model (BM) canvas as a framework. In the context of the CEBMs reviewed, the foundational BM was adapted to align with CE principles. This adaptation involved integrating CE principles into the core elements of the BM to minimise the depletion of natural resources, allowing companies to maintain competitive advantage while fostering sustainability and environmental stewardship.

4.1 Introduction

Understanding how the concept of the circular economy (CE) transitions from theory to practical application begins with examining its origins and significance. A relevant example of this theoretical and practical overlap is the business model, which describes how a company creates, delivers, and captures value [57, 69]. The term "business model" first appeared in academic literature in 1957 by Bellman et al. [7]. The study explored the creation of business games for training purposes. The term was mentioned only once in the study, in the sentence: "And many more problems arise to plague us in the construction of these business models than ever confronted an engineer" (p. 474). This initial usage linked the business model concept to replicating reality through a model. The term resurfaced in an academic context in Jones

© The Author(s), under exclusive license to Springer Nature Switzerland AG 2025
O. Otasowie et al., *Circular Economy Business Model for Construction Organisations*, Green Energy and Technology, https://doi.org/10.1007/978-3-031-88322-4_4

[40], which addressed how business students should be educated and how emerging technologies should be introduced. Interestingly, the phrase "business model" did not appear within the text, suggesting it was chosen arbitrarily for the title. Its origins, therefore, seem tied to simplifying complex realities to teach future managers about technology.

For decades, the term was rarely used in scholarly work. A review by Osterwalder et al. [58] highlights that, until the 1990s, only a handful of peer-reviewed journal articles included "business model" in their titles, with just five papers during the entire decade. However, as information and communication technologies (ICT) and Internet-based businesses emerged, the term gained traction among academics and practitioners alike. During this period, the phrase "business model" also became associated with the NASDAQ index's trends, especially during the rise and subsequent burst of the dot-com bubble. From the 1990s onward, the concept began to permeate various disciplines, including marketing, management, finance, and ICT. It was applied within various frameworks [25], this further cemented its importance in academic and professional discourse.

Zott et al. [80] suggest that business models involve activities to generate and capture value. In contrast, Teece [69: 172] views a business model as

> the design or architecture of the value creation, delivery and capture mechanisms employed. The essence of a business model is that it crystallises client needs and ability to pay, defines how the business enterprise responds to and delivers value to clients, entices clients to pay for value, and converts those payments to profit through the proper design and operation of the various elements of the value chain.

Osterwalder and Pigneur [57: 14] described a business model as "the rationale of how an organisation creates, delivers, and captures value." Building on this concept, the study introduced a widely recognised framework called the Business Model Canvas (Fig. 4.1). Over time, this canvas has become a popular tool for illustrating and communicating the structure of nine critical components: key partners, key resources, key activities, value propositions, client relationships, channels, client segments, revenue streams, and cost structures.

4.2 Theoretical Underpinnings of Circular Economy Business Models

Recent studies highlight the limited unified theory or conceptual framework for implementing CE principles. For example, Fischer and Pascucci [24: 5] note that "the identification of relevant strategic processes and actors in this domain is still limited." To address this gap, this book explores the rationale for adopting innovative Circular Economy Business Models (CEBMs) or adapting existing ones, providing theoretical insights by integrating perspectives from different fields. It draws on neo-institutional theory [14] from institutional theory and the natural-resource-based view [30] from strategic management. These approaches align with resource-based

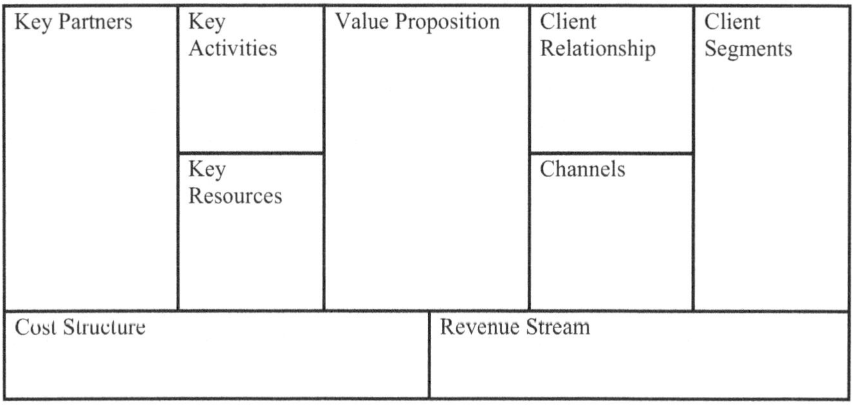

Fig. 4.1 Business model canvas. [57] *Source* Osterwalder and Pigneur

and institutional logic, widely used in Business and Natural Environment research [28, 47, 53].

4.2.1 Natural-Resource-Based-View of the Firm

The resource-based view (RBV) explores how firms can sustain long-term competitive advantages by identifying their roots within strategic management theory [4]. According to this perspective, organisations can secure and maintain a competitive edge when they possess resources and capabilities that are valuable, rare, difficult to imitate, and lacking in viable substitutes, assuming resource heterogeneity. Such valuable resources enable firms to leverage internal strengths to capitalise on external opportunities or mitigate weaknesses and threats from the external environment. Rare resources are considered specific to a firm, originating from unique combinations of factors within the business. These resources are not easily replicable by competitors. Barney [4] highlights that resources often attain their uniqueness due to characteristics like ambiguity, social complexity, or tacit knowledge. Tacit resources arise from practical experience and expertise, making them invisible and challenging for competitors to replicate. Casual ambiguity further enhances inimitability by obscuring the connection between a firm's resources and its competitive advantage.

Socially complex resources stem from dynamic interactions between various organisational elements working collectively toward shared objectives. Given the intricate ways firms structure themselves to harness strengths and mitigate risks, these resources are also problematic for competitors to understand or duplicate. In addition, resources that cannot be replaced with alternatives by competitors are considered non-substitutable, further solidifying a firm's competitive position. However, Hart [30] criticised Barney's model for overlooking the reality of finite natural resources and the impact this has on the competitiveness of businesses. In response, Hart proposed the firm's natural resource-based view (NRBV), emphasising that engaging in environmentally sustainable practices is essential for achieving a durable competitive edge. He identified three critical capabilities within this framework: pollution prevention, product stewardship, and sustainable development. *Pollution prevention*, which minimises waste and resource use during the manufacturing phase of a product's life cycle, can lower costs by improving resource efficiency and reducing waste management and compliance expenses [30]. *Product stewardship*, on the other hand, seeks to mitigate environmental harm throughout the product's entire life cycle. Tools such as life cycle assessment and eco-design enable better evaluations of environmental impacts and drive the creation of green products. In this context, Hart [30: 994] argued that competitive advantage stems from "*competitive preemption*," such as securing limited resources or setting new industry standards, rather than solely from cost reduction.

Sustainable development represents the third pillar, addressing social and environmental challenges simultaneously. This approach extends beyond pollution reduction, aiming to create inclusive products for underserved markets in developing regions. Here, competitive advantage arises from innovation and new market opportunities. Hart's NRBV remains highly relevant considering today's ecological crises. The deteriorating state of natural ecosystems directly and indirectly affects organisational management and long-term viability. Hart's model also highlights the rationale for adopting circular economy business models (CEBMs). With natural capital declining [76], increasing global influence [43], and the rising costs and volatility of resources affecting linear business models [19], businesses face mounting pressures. Also, stricter climate and waste regulations and societal expectations for companies to address the environmental problems they contribute to [19, 29] underscore the urgency of change.

In this context, business model innovation aimed at circularity can offer significant advantages, such as enhanced brand reputation, stronger customer loyalty, new revenue opportunities, reduced material costs and supply risks, and compliance with regulations. These benefits collectively represent what Lacy and Rutqvist [43] describe as a "circular advantage." The NRBV is a significant development within this theoretical framework [5] and has garnered substantial attention in academic discussions [61], as thoroughly reviewed in the literature [33, 53]. Nevertheless, Hart and Dowell [33: 1475] emphasised the "need for research that continues to evaluate and extend the propositions offered by Hart [30]." Hart's propositions can be further developed through the lens of innovation capabilities within CEBM. This is particularly relevant for two key reasons. First, the evolving competitive environment presents firms with new opportunities to establish a circular advantage. Second, Hart's framework and related studies seem to either insufficiently address or only implicitly consider the role of business model innovation in responding to ecological and societal challenges. For example, it is argued that:

> sustainable economies and corporations (...) cannot be based on continuing growth in the consumption of non-renewable energy and virgin raw materials. Nor can they create

> hazardous waste and polluting emissions. Environmental sustainability requires completely redesigning organisations and strategies [66: 157].

While business models (BMs) are not explicitly mentioned, it can be inferred from the statement that "the full redesign of organisations" would encompass BMs as a critical component. Hart and Milstein [34] subtly acknowledge the potential for a shift in BMs, identifying 'sustainability' as a transformative force in business. They argue that "in the long run, however, the dynamics of creative destruction will work against firms that rely only on incremental improvements and fail to change the fundamental manner in which they provide products, processes, and services" (p. 24). In contrast, Hart [31, 32] and earlier study, Hart and Milstein [34] provide a more explicit acknowledgement of BM evolution. Their research focuses on strategies targeting the "bottom of the pyramid," emphasising that traditional BMs designed for affluent markets are inadequate. Instead, they advocate for a fundamental redesign of BMs to deliver affordable goods and services to underserved populations.

Moreover, efforts to reconnect ecological and economic systems have been proposed [23]. CEBMs align with the firm's NRBV and support Hart and Dowell's [33] call for management scholars and practitioners to move beyond incremental approaches. These approaches should not only aim to minimise harm, such as eco-efficiency, product stewardship, and pollution prevention, but should also directly address broader social and environmental challenges. The cooperative nature required for value generation in CEBMs supports the validity of Hart [30] ideas, as highlighted by Webster [74]. While resource-based theories often prioritise competition over collaboration [29, 68], Hart [30] wisely noted that a firm's survival relies not only on competition but also on social legitimacy, which opens the door for collaboration to achieve this legitimacy. This view aligns with the essential role of cooperation in CEBM activity systems. Moreover, the broader relationship structures within CEBMs could address some of the criticisms initially directed at the NRBV of the firm, particularly its tendency to consider firms in isolation, which has been critiqued as overly atomistic [45].

While the literature on BMs concurs with resource-based theories, which suggest that economic value creation potential grows as the resources and capabilities supporting the BM become valuable, rare, and difficult to imitate [1], the role of resource control, uniqueness, causal ambiguity, and social complexity takes on a more nuanced character in circular approaches. This shift corroborates Dyer and Singh [17] relational perspective, which posits that competitive advantages may extend beyond one organisation. As a result, the interaction between the NRBV and CEBMs may lead to new insights, suggesting that the NRBV remains a relevant theoretical framework for understanding the justification for adopting CEBMs. In this way, CEBMs might expand and refine Hart's model and some of its assumptions.

4.2.2 *Institutions and Neo-Institution Theory*

According to Douglass North, a Nobel Prize-winning economist, institutions are "the rules of the game in a society or, more formally, (...) the humanly devised constraints that shape human interaction" [55: 3]. In this framework, institutions serve to establish boundaries for decision-making, thereby reducing uncertainty and minimising the transaction costs involved in individuals pursuing their goals. On the other hand, the social constructivist perspective, as articulated by scholars such as Vatn [73], suggests that societal structures play a significant role in shaping individual preferences and decisions. This approach posits that "individuals interact to form institutions, while individual purposes or preferences also are moulded by socioeconomic conditions. The individual is both a producer and a product of her circumstances" [37: 177]. Sociologists, such as Scott [63: 33], contribute to this discussion by categorising institutions into three types: "cognitive, normative, and regulative structures and activities that provide stability and meaning to social behaviour". Regulative institutions are formal rules, while normative institutions represent values and standards that define appropriate behaviour [15, 64]. Cognitive institutions, meanwhile, involve "shared conceptions that constitute the nature of social reality and the frames through which meaning is made" [64: 57].

In organisational studies, institutions have been a focal point of research, mainly through the lens of neo-institutional theory [14, 52, 56]. Di Maggio and Powell [14] argue that organisational activities are influenced by the broader "organisational field", a network of organisations that interact and shape each other. This framework highlights three key forces: normative, coercive, and mimetic. Normative pressures, often arising from universities and professional networks, define acceptable professional and organisational conduct. Mimetic forces drive organisations to imitate successful strategies in uncertain environments. Coercive pressures, on the other hand, typically stem from regulatory entities such as government agencies. Scott [63: 56] further elaborates on the organisational field, describing it as "a community of organisations that partake of a common meaning system and whose participants interact more frequently and fatefully with one another than with actors outside the field". In his opinion, the field includes any player influencing organisations in a regulatory, normative, or cognitive way. These forces ultimately result in organisational "homogeneity in structure, culture and output" [14: 147] and "the concept that best captures the process of homogenisation is isomorphism" (p. 149). According to the widely held belief that fields develop, they gather around similar goods, markets, or technology. However, Hoffman [38: 352] also suggested that fields grow "around the issues that become important to the interests and objectives of a specific collective of organisations". Businesses' ability to secure resources improves when they align with institutional expectations, as this ensures legitimacy and social approval, both of which are crucial for maintaining a competitive advantage [14, 56].

The institutional entrepreneurship framework [56] introduces a focus on agency within the neo-institutional theory, enabling organisations to react to institutional pressures in ways that go beyond mere compliance [8]. While this perspective highlights diversity and variation, particularly in the form of differing organisational changes, the meso-level viewpoint of neo-institutional theory emphasises the spread of organisational forms and practices. This diffusion process underscores homogeneity and convergence within organisational fields [10, 35]. Integrating agency and rational decision-making with institutional pressures brings a dynamic dimension to studying institutional environments [13, 35, 53].

The neo-institutional theory has been extensively applied in corporate sustainability research, providing insights into how broader institutional contexts influence organisational environmental practices [27, 48]. This theoretical lens is particularly suitable for understanding the rationale behind adopting CEBMs. Organisations often conform to the norms of their organisational fields to achieve legitimacy and societal acceptance, which can subsequently enhance competitive advantage. Organisations may gain legitimacy and improve competitiveness by aligning their business model strategies with institutional expectations emerging from discussions around the circular economy. These field structures are characterised by active engagement and interaction among multiple organisations within a shared discourse [14]. Although it is still too soon to identify a well-structured organisational framework addressing the issue of CE, specific societal processes appear to support its development. With efforts led by the Ellen MacArthur Foundation (EMF), the CE concept has recently gained attention. This is with a mission "to facilitate the transition towards the CE" [23], the EMF, working in partnership with the World Economic Forum, McKinsey & Company, SUN, and SYSTEMIQ, has published several reports advocating for the economic rationale behind the CE transition. These initiatives have sparked dialogue among organisations, policymakers, and academic institutions. Projects like the New Plastics Economy and the Circular Fibres Initiative have started to outline strategies for shifting the global textiles and plastics sectors from a linear model to a circular one. Furthermore, other organisations, including the European Union with its CE Action Plan, have also played a critical role in advancing regulatory reforms, prompting corporate leaders and policymakers globally to adopt circular principles.

Hence, neo-institutional theory offers valuable insights while presenting limitations in the context of research on CEBMs. Traditionally, this theory views the natural environment merely as a source of disruptions that drive organisational change [68]. However, it often overlooks the integration of organisations within ecological systems, treating the environment as separate from organisational domains. Hoffman and Jennings [39: 20] echo this concern, arguing that "interests of the natural environment (and future generations) are not captured within standard institutional analyses, which are social and present by definition." Applying CE principles and designing business models inspired by biological systems could reshape these fields, positioning nature as a source of shocks and a model organisations can learn from. This perspective aligns with stakeholder theory, which emphasises the need to recognise nature as a legitimate stakeholder directly affected by its influence on business activities [16, 67, 75]. Viewing nature as a stakeholder reinforces its central role in shaping organisational strategies and highlights the interdependence between business and ecological systems. Starik [67: 212] provided the following advice:

> Rather than overly restricting the number of natural environment stakeholders, the continued human-caused environmental deterioration of the planet appears to call for all organisations to consider as stakeholders as many natural environment entities as possible (...). Adding non-human natural environment stakeholders could make an organisation's stakeholders' map more nearly complete for total environmental problem identification, analysis, evaluation, and resolution.

The development of the field would align with the trajectories envisioned by Hoffman and Jennings [39] and Starik and Kanashiro [68]. While this evolution has implications, it underscores the inherent paradoxes within institutional theories. Without deliberate intervention, the environment would remain excluded from the field. Integrating nature as a source of learning requires a fundamental shift in how the environment is institutionalised within laws, traditions, and belief systems. This shift may also necessitate significant de-institutionalisation, affirming Randles and Laasch [60: 60] observation that "any process of institutionalisation must involve a corresponding process of de-institutionalisation." Two concepts from institutional theory are evident in the institutional transformations associated with the CE. Initially, the EMF and pioneering business leaders can be understood as institutional entrepreneurs, defined by Maguire et al. [50: 657] as "actors who have an interest in particular institutional arrangements and who leverage resources to create new institutions or to transform existing ones", driven by changing socioeconomic, technological, and ecological contexts. Subsequently, as CE principles become more entrenched at a societal level, emerging regulatory, normative, and mimetic pressures may lead to organisational and institutional isomorphism. This dual phenomenon where entrepreneurial innovation and institutional isomorphism around CE co-occur highlights the contradictory aspects of institutional theories. On the one hand, these theories emphasise change, on the other, they promote conformity [26]. Thus, the concept of fields becomes essential to understanding how organisational conformity evolves and ultimately achieves legitimacy within neo-institutional theory.

4.3 Evaluating Circular Economy Business Models

This section examines research on CEBMs and the theoretical framework underpinning these studies. Within CE, the focus on creating new business models or adapting existing ones is logical. Adopting CE principles impacts all facets of a business model, including value proposition, value creation and delivery, and value capture. For instance, products with medium-to-long lifespans, such as household appliances, do not always need to be purchased outright under circular production and consumption systems. Instead, they can be leased or accessed through pay-for-use arrangements. Under these models, consumers pay to use the product over a specified period, with payments linked to performance [22, 43]. This structure motivates manufacturers to design durable products, as they retain ownership and are responsible for maintenance [36]. In addition, when products are returned at the end of their lifecycle, manufacturers can recover secondary raw materials. Long-term customer relationships also enable producers to reduce dependency on costly raw materials while benefiting from a steady revenue stream without requiring customers to invest heavily in ownership. This approach introduces a new value proposition, such as prioritising access over ownership to enhance customer value. Similarly, it reshapes value creation and delivery by requiring maintenance, repair, and customer relationship management competencies. The value capture aspect also changes, as income is generated from services rather than product sales, potentially lowering costs. For example, Bundles, a Dutch start-up, offers a service model where customers pay for the use of innovative, internet-connected washing machines based on the number of washes [20]. These machines are designed to be durable and recyclable, ensuring they can be returned, repaired, and refurbished to start a new use cycle.

Although early CE literature emphasises the importance of innovative business models, it often provides only broad insights into their characteristics. However, performance-based payment models, as opposed to conventional consumer ownership models, are essential for promoting durable and reusable product designs [22].

4.3.1 Framework Underpinning Circular Economy Business Models

Different frameworks outline CE principles from which CEBMs can be derived. One such framework is DISRUPT, which emphasises principles such as designing for the future, integrating digital technologies, sustaining and preserving resources, rethinking business models, using waste as a resource, and prioritising regenerative resources. However, the most widely cited framework in the literature is ReSOLVE, developed by the Ellen MacArthur Foundation [23]. This framework identifies six key actions for implementing CE principles: Regenerate, Share, Optimize, Loop, Virtualise, and Exchange. The principle of regeneration focuses on ensuring that biological resources extracted from ecosystems are replenished and encourages using renewable energy. Sharing maximises product utilisation by facilitating shared or sequential use, promoting reuse. Optimisation seeks to enhance product efficiency while minimising waste throughout the production process. Looping ensures materials remain within a closed system, extending their lifecycle. The virtualisation principle emphasises delivering value through digital or virtual means instead of physical materials. Finally, exchange focuses on replacing traditional materials and technologies with innovative alternatives. Table 4.1 organises various CEBMs from the literature according to the ReSOLVE framework.

Table 4.1 Circular economy business model types

S/N	CEBM	Source	Business focus/Opportunity
1	Energy recovery	Damen [11], Lacy et al. [41]	**Regenerate** (transforming waste materials that cannot be recycled into usable energy, such as heat or electricity)
2	Circular suppliers	Lacy et al. [41], EMF [19]	**Regenerate** (Harnessing sustainable energy sources)
3	Efficient buildings	Scot (2015)	**Regenerate** (Positioning business operations within energy-efficient facilities)
4	Sustainable product locations	Scot (2015)	**Regenerate** (Establishing businesses within eco-industrial parks)
5	Chemical leasing	Moser and Jakl [54]	**Regenerate** (Selling only functions performed by the chemical)
6	Maintenance and repairs	Lacy et al. [41], Bakker et al. [3], WRAP [78], Planing [59], Damen [11]	**Sharing** (Extending product life through maintenance and repair)
7	Collaborative consumption, Sharing platforms, PSS: Product renting, sharing or pooling	Lacy et al. [41], Lacy et al. [42], WRAP [78], Planing [59], Tukker [70], de Jong et al. [12]	**Sharing** (Facilitating the shared utilisation, access, or ownership of a product among individuals or between organisations.)
8	PSS: Product lease	Tukker [70], Jong et al. [12], WRAP [77]	**Sharing** (Use a product exclusively without being the owner)
9	PSS: Availability based	Van Ostaeyen et al. [71], Mentink [51]	**Sharing** (The product or service being available for the client for a specific period)
10	PSS: Performance-based	Van Ostaeyen et al. [71], Zairul et al. [79]	**Sharing** (Revenue is earned based on the provided solution, the achieved outcome, or the fulfilment of a specific demand)
11	Incentivised return and reuse or Next Life Sales	WRAP [77], Mentink [51], Lacy et al. (2022); Damen [11]	**Sharing** (Customers bringing back previously used items in exchange for a predetermined value)
12	Upgrading	Planing [59], Mentink [51]	**Sharing** (Upgrading components to higher-quality alternatives)
13	Product attachment and trust	Mentink [51]	**Sharing** (Designing products that are cherished and built to endure)
14	Bring your device	WRAP [77]	**Sharing** (User bringing their devices to get access to services)
15	Hybrid model	Bakker et al. [3]	**Sharing** (A durable product containing short-lived consumables)

(continued)

Table 4.1 (continued)

S/N	CEBM	Source	Business focus/Opportunity
16	Gap-exploiter model	Bakker et al. [3], Mentink [51]	**Sharing** (Exploiting "lifetime value gaps" or leftover value in product systems)
17	Asset management	WRAP [77]	**Optimise** (Internal collecting, reusing, refurbishing and reselling of used products)
18	Produce on demand	van Renswoude et al. [72], WRAP [77], Scott [62]	**Optimise** (Manufacturing only when there is a confirmed demand and an existing order)
19	Good housekeeping, Waste reduction, Fit thinking, Lean thinking,	van Renswoude et al. [72], Scott [62], El-Haggar [18], Bautista-Lazo [6]	**Optimise** (Minimising waste during the manufacturing process)
20	PSS: Activity management/ outsourcing	Tukker [70]	**Optimise** (Leveraging capital assets, materials, and labour more effectively by outsourcing certain functions or services)
21	Remanufacturing, product transformation	Damen [11], Planing [59], Lacy et al. [42]	**Loop** (Reconditioning a product or its parts to a state equivalent to new)
22	Recycling, Recycling 2.0, Resource recovery	Lacy et al. [42], Damen [11], Planing [59], Lacy et al. [41]	**Loop** (Extracting valuable materials from discarded products or waste byproducts)
23	Upcycling	Lacy et al. [41], Mentink [51], Planing [59]	**Loop** (Repurposing materials and enhancing their value)
24	Circular suppliers	van Renswoude et al. [72], Lacy et al. [41]	**Loop** (Utilising materials from closed-loopsystems and bio-based or entirely recyclable resources)
25	Dematerialised services	WRAP [77], van Renswoude et al. [72]	**Virtualise** (Transforming tangible products, services, or processes into digital or virtual formats)
26	New technology	EMF [19]	**Exchange** (Utilising innovative production technologies)

Adapted from Lewandowski [44]

Limited research is available in the literature on CEBMs. Linder and Williander [46: 2] define a CEBM as:

> a business model in which the conceptual logic for value creation is based on utilising economic value retained in products after use in the production of new offerings. Thus, a circular business model implies a return flow to the producer from users, though there can be intermediaries between the two parties. Therefore, the term 'circular business model' overlaps with the concept of closed-loop supply chains and always involves recycling, remanufacturing, reuse or one of their sibling activities (e.g. refurbishment, renovation, and repair).

Hence, CEBMs are recognised as mechanisms for generating value by recycling resources and materials traditionally viewed as waste at the end of their lifecycle. To conceptualise a CEBM, adopting a business model framework from literature is necessary. The widely used framework developed by Osterwalder and Pigneur [57], which consists of nine dimensions, has been applied in various studies [21, 44] within the emerging CE literature to develop CEBM canvases.

4.3.2 Lewandowski's Circular Economy Business Model Canvas

Lewandowski [44] introduced a framework for CEBMs, building on the core principles of the CE outlined in the ReSOLVE framework. This framework expanded upon Osterwalder and Pigneur's [57] BM canvas by incorporating additional components, specifically take-back systems and adoption factors. According to Lewandowski [44], the primary elements of the CEBM canvas include value proposition, customer segments, channels, customer relationships, revenue streams, key resources, key activities, key partnerships, cost structure, take-back systems, and adoption factors. The framework emphasises the importance of material loops, which the study identified as central to CE principles. It assumes that products and their components and materials can be collected through reverse logistics processes and either cascaded (for biological materials), reused, redistributed, remanufactured, refurbished, or recycled (for technical materials). These reverse logistics processes align with CE concepts by focusing on take-back systems, incentivised returns and reuse, and the recovery of used products. Lewandowski [44] also highlighted organisations' need to understand and address potential criticisms of CEBMs to ensure effective implementation. Transitioning to a CE-based business model depends on various factors, including internal capabilities and external conditions. Internally, the organisation's ability to adopt CE principles is crucial to successfully transitioning its business model.

In summary, Lewandowski‘s CE Business Model Canvas (BMC) expands the concept of "Value Proposition" in CE by integrating ideas such as "Product Service Systems (PSS)," "Circular Products," "Virtual Services," and "Client Incentives within Take Back Systems (TBS)." These concepts are drawn from the Ellen MacArthur Foundation's ReSOLVE framework. In addition, the circular BMC introduces two new components beyond the traditional nine building blocks: the "Take Back System" and "Adoption Factors." The "Take Back System" specifically addresses developing a closed-loop supply chain or reverse logistics channel, facilitating material recovery and reuse.

4.4 EMF and IDEO's Circular Economy Business Model Canvas

The framework for a CEBM proposed by the Ellen MacArthur Foundation (EMF) and IDEO [21] integrates the core principles of the CE as outlined in the ReSOLVE framework. This framework is presented as a CEBM canvas, adapted from Osterwalder and Pigneur's [57] BM canvas. It includes key components such as value proposition, customer segments, channels, customer relationships, revenue streams, key resources, key activities, key partnerships, and cost structure. The CEBM canvas emphasises a shift in organisational value propositions from selling physical products to offering services, promoting digital solutions, and minimising waste through strong and sustainable relationships with users and customers. This approach bridges the principles of the circular economy, as outlined in the ReSOLVE framework, with the structure of Osterwalder and Pigneur's [57] BM canvas to encourage the development of sustainable and circular business practices.

In summary, the CEBM canvas, developed by EMF and IDEO, is inspired by the core principles outlined in the ReSOLVE framework for the CE. This canvas presents a series of reflective questions to guide the design process of a CEBM. It is structured on the foundation of Osterwalder and Pigneur's [57] BM Canvas.

4.4.1 Ludeke-Freund et al.'s Circular Economy Business Model Canvas

Lüdeke-Freund et al. [49] also introduced a framework for CEBMs inspired by the core principles of the CE outlined in the ReSOLVE framework. This CEBM canvas emphasises organisational value propositions shifting from selling physical products to offering services, delivering digital products, minimising waste, and driving substantial, ongoing relationships with users and clients. The model integrates the key principles of the ReSOLVE framework with the structure of Osterwalder and Pigneur's [57] Business Model Canvas.

In summary, Lüdeke-Freund et al. [49] developed the CEBM canvas by incorporating the core principles of the CE outlined in the ReSOLVE framework. This canvas includes prompts to encourage critical thinking about the structure and design of a CEBM. Its foundation draws inspiration from the business model canvas Osterwalder and Pigneur [57] created.

4.4.2 Antikainen and Valkokari's Circular Economy Business Model Canvas

Antikainen and Valkokari [2] proposed a framework for CEBM based on the core principles outlined in the ReSOLVE framework. Their approach integrates drivers, stakeholder engagement, and sustainable impacts while incorporating elements from Osterwalder and Pigneur's [57] Business Model Canvas. In summary, Antikainen and Valkokari [2] extend the scope of the traditional BMC by adopting a more comprehensive perspective that incorporates multiple levels of analysis. Their approach introduces additional elements such as the "Business Ecosystem Level," "Sustainability Impact," and iterative cycles for evaluating "Sustainability and circularity" within the business model. This framework reflects systemic-level changes and emphasises the need for a holistic approach when developing circular business models. The study highlights the significance of "Trends and Drivers" in driving systemic innovation and underlines the importance of involving stakeholders at the "Firm Ecosystem Level." This includes internal and external participants such as network partners, consumers, suppliers, society, and the environment, all of whom play a critical role in shaping and influencing the business.

These models offer both benefits and limitations, which the authors recognise. For instance, Lewandowski's [44] canvas has raised questions about its practical applicability, as it relies exclusively on a literature review and lacks empirical evidence to support its findings. In contrast, the framework introduced by Antikainen and Valkokari [2] has been applied to a real-world context involving a Finnish social enterprise. The authors observed that while the framework was practical due to its simplicity, gathering and integrating the required information was time-intensive.

4.4.3 Measuring Circular Economy Business Models

The CEBMs discussed highlight that a business model (BM) outlines how value is created, delivered, and captured. It is widely recognised as a key driver of competitive advantage, as it aligns with customer needs and willingness to pay. Notably, the core components of BMs form the foundation of all reviewed CEBMs. Identifying and organising these elements provides organisations with the framework to create value, achieve profitability, and sustain success. However, to align with CE principles, traditional BMs were adapted by integrating these principles into their core elements. This adaptation minimises the wasteful exploitation of natural resources, enabling companies to maintain a competitive edge while driving a sustainable relationship with the environment.

Several models, such as those by Lüdeke-Freund et al. [49], the EMF and IDEO [21], and Antikainen and Valkokari [2], applied CE principles to BM elements.

Among these, Lewandowski's [44] model is particularly noteworthy for its comprehensive approach. It emphasises a closed-loop production process, where materials or products at the end of their lifecycle are recovered through a take-back system. These materials can then be reused in new production cycles or, if reuse is not feasible, composted to provide environmental nutrients. Given its focus on closed-loop processes and resource recovery, this study adopted Lewandowski's [44] framework as the theoretical foundation.

Analysing the existing frameworks for CEBMs, the key components influencing CE principles in organisational business models include client segments, value propositions, channels, client relationships, revenue stream, key resources, key activities, key partnerships, cost structure, and take-back systems. This study examines these ten components while assessing the significance of two additional factors: materials management systems and organisational culture, specifically within the construction industry. Including these new factors addresses ongoing debates among researchers regarding the relative importance of the original ten variables. To evaluate CEBMs in construction organisations, this study incorporates the foundational elements shaping CEBMs and the newly proposed constructs.

4.5 Developing CEBMS for the Construction Industry

It is essential to recognise that while organisations within the same industry may adopt CEBMs with similar components, these models are unique to each firm and not representative of an entire industry. CEBMs outline specific market focuses and ventures, and they are inherently dynamic, adapting over time to align with the evolving business environment. Developing tailored CEBMs is crucial for successfully integrating CE principles in the construction sector. As previously noted, adopting CE in construction can significantly reduce the industry's environmental footprint, which has been growing steadily. The CE concept emphasises the efficient use of materials and resources. Implementing CE principles generates substantial financial savings, amounting to hundreds of billions of US dollars, while mitigating the adverse impacts of human activities on the environment [44]. The ReSOLVE framework from the Ellen MacArthur Foundation incorporates CE principles, offering diverse perspectives on circular business practices. According to Planing [59], transitioning from a linear to a CE within the construction sector necessitates four foundational elements: designing products and materials for circularity, rethinking or transforming business models, establishing global reverse logistics networks, and driving an enabling environment.

Lewandowski [44] highlighted that transitioning to CE principles might require creating new business models or adapting existing ones. Transforming business models involves integrating CE principles into their core components, aiming to eliminate the inefficient use of natural resources. Businesses are encouraged to adopt circular processes that provide a competitive advantage while maintaining

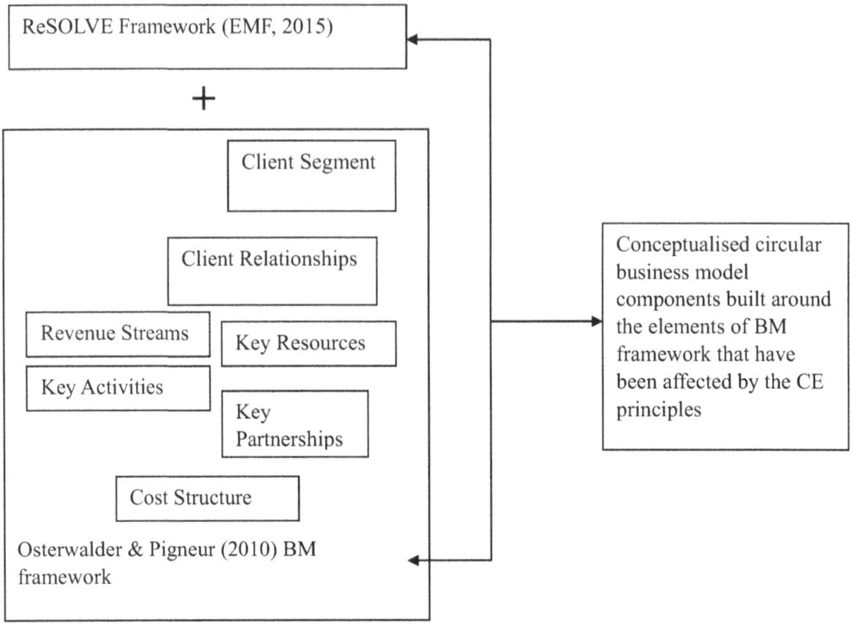

Fig. 4.2 Author's CEBM conceptualisation process

harmony with the natural environment. This transition entails implementing closed-loop production systems where materials or products that reach the end of their usefulness are recovered for reuse in new production cycles. If reuse is not feasible, materials can be composited to return nutrients to the environment. To create effective and efficient CEBMs for the construction industry, CE principles must be integrated into every business model element. Figure 4.2 illustrates the conceptualisation process of CEBMs.

While the construction industry has the potential to adopt new mechanisms, concepts, frameworks, and models, many of these innovations are developed primarily for other sectors, such as manufacturing. Consequently, these frameworks may not fully address the unique requirements of the construction sector. This limitation is particularly evident in the context of CEBMs. However, this gap can be bridged during implementation by incorporating additional elements tailored to the specific needs of the construction industry. For instance, Lewandowski [44] introduced two additional elements (take-back systems and adoption factors) into the existing CEBM framework, recognising that the original components were insufficient to achieve full circularity. Similarly, Antikainen and Valkokari [2] added drivers, stakeholder involvement, and sustainability impact to enhance the CEBM framework.

Based on the literature, it is evident that developing an effective CEBM for construction organisations requires a framework. Adding relevant CEBM elements to address construction-specific challenges is a critical first step. According to Sequeira

[65], conceptualisation involves breaking down and translating ideas into standard terms to establish a shared understanding among stakeholders. This process combines an idea's essential characteristics and specifics, forming a foundation for creating a coherent framework. Thus, conceptualising a circular business framework is crucial in developing CEBMs for the construction industry. The final phase in this process involves designing CEBMs based on the conceptual framework. Osterwalder and Pigneur [57] outline five stages in a business model design process: understanding, mobilising, designing, implementing, and managing. In addition, De Jong et al. [12] propose steps for designing CEBMs, starting with understanding and learning CE principles and progressing to designing models tailored to the services an organisation can deliver while considering client expectations and willingness to pay.

A well-conceptualised framework can generate various CEBM designs tailored to the construction industry, presenting multiple business opportunities. These designs may focus on product or material-driven strategies, product performance, building performance, or collaboration-driven models. Regardless of the chosen approach, these designs aim to achieve economic growth without depleting natural resources, thereby avoiding the environmental degradation typically associated with extracting virgin materials.

4.5.1 Measuring Value in Circular Economy Business Models

The conceptual foundation of the CEBM is most effectively structured around the concept of value. Within CE literature, value is a central theme, described as "an economy that provides multiple value creation mechanisms which are decoupled from the consumption of finite resources" [23, p. 23]. Similarly, value plays a pivotal role in business model (BM) literature, where the BM framework revolves around value, as illustrated in models such as Osterwalder and Pigneur [57]. By analysing the defining aspects of value propositions, value creation and delivery, and value capture, the CEBM can be conceptualised by integrating the value-oriented dimensions of the BM framework with the impact of CE principles on these elements. A valuable tool in this context is the ReSoLVE framework [23]. This framework consolidates various CE-related strategies (Regenerate, Share, Optimize, Loop, Virtualise, and Exchange) under a unified structure, providing practical guidance for implementing CE principles effectively.

Determining value often involves connecting client jobs with products and services through one of four key components, a practice that has gained widespread acceptance [9]:

1. Value Enhancers: Outline the advantages or unexpected benefits that specific products or services offer to clients.

2. Problem Solvers: Detail how products and services alleviate clients' challenges or frustrations with current alternatives.
3. Client Objectives: Highlight the outcomes or goals consumers aim to achieve by completing a specific task.
4. Client Challenges: Identify the difficulties clients experience with current methods, including concerns about cost, time investment, or barriers to task completion.

4.6 Summary

This chapter analyses the existing literature on CEBMs. The findings reveal that existing models were developed based on BM core elements and subsequently transformed to incorporate CE principles. However, Lewandowski's CEBM stands out as the most comprehensive model. It emphasises adopting a closed-loop production process, in which materials or products at the end of their lifecycle are recovered through take-back systems. These recovered materials can either be reused as inputs for new production processes or, if reuse is unfeasible, composted to enrich the environment with nutrients. In addition, the review identifies key determinants of CEBMs, including value proposition, customer segments, channels, customer relationships, revenue streams, key resources, key activities, key partnerships, cost structure, and take-back systems. These elements collectively underpin the development and implementation of effective CEBMs. Lastly, the gaps revealed by examining existing CEBMs are explored in the next chapter.

References

1. Amit R, Zott C (2001) Value creation in e-business. Strateg Manag J 22(6–7):493–520
2. Antikainen M, Valkokari K (2016) Framework for sustainable circular business model innovation. In: ISPIM innovation symposium, p 1. The International Society for Professional Innovation Management (ISPIM)
3. Bakker CA, Den Hollander MC, Van Hinte E, Zijlstra Y (2014) Products that last: product design for circular business models. TU Delft Library
4. Barney J (1991) Firm resources and sustained competitive advantage. J Manag 17(1):99–120
5. Barney JB, Ketchen DJ Jr, Wright M (2011) The future of resource-based theory: revitalization or decline? J Manag 37(5):1299–1315
6. Bautista Lazo S (2013) Sustainable manufacturing: turning waste into profitable coproducts. https://livrepository.liverpool.ac.uk/12933/
7. Bellman R, Clark CE, Malcolm DG, Craft CJ, Ricciardi FM (1957) On the construction of a multi-stage, multi-person business game. Oper Res 5(4):469–503
8. Boxenbaum E, Jonsson S (2008) Isomorphism, diffusion and decoupling. In: Greenwood R, Oliver C, Suddaby R, Sahlin K (eds) The SAGE handbook of organizational institutionalism. Sage, pp 78–99
9. B2B International (2023) What is the value proposition canvas. Retrieved from https://www.b2binternational.com/research/methods/faq/what-is-the-valueproposition-canvas/

10. D'Aunno T, Succi M, Alexander JA (2000) The role of institutional and market forces in divergent organizational change. Adm Sci Q 45(4):679–703. https://doi.org/10.2307/2667016
11. Damen MA (2012) A resources passport for a circular economy. https://studenttheses.uu.nl/bitstream/handle/20.500.12932/12157/A%20resources%20passport%20for%20a%20circular%20economy.pdf?sequence=1
12. De Jong E, Engelaer F, Mendoza M (2015) Realizing opportunities of a circular business model. http://circulatenews.org/2015/04/de-lage-landen-realising-theopportunities-of-a-circular-business-model
13. Delmas MA, Toffel MW (2010) Institutional pressures and organizational characteristics: implications for environmental strategy. Harvard Business School Technology & Operations Mgt. Unit Working Paper, (11-050)
14. DiMaggio PJ, Powell WW (1983) The iron cage revisited: institutional isomorphism and collective rationality in organizational fields. Am Sociol Rev 48(2):147–160
15. Doh JP, Howton SD, Howton SW, Siegel DS (2010) Does the market respond to an endorsement of social responsibility? The role of institutions, information, and legitimacy. J Manag 36(6):1461–1485
16. Driscoll C, Starik M (2004) The primordial stakeholder: advancing the conceptual consideration of stakeholder status for the natural environment. J Bus Ethics 49:55–73
17. Dyer JH, Singh H (1998) The relational view: cooperative strategy and sources of interorganizational competitive advantage. Acad Manag Rev 23(4):660–679
18. El Haggar S (2010) Sustainable industrial design and waste management: cradle-to-cradle for sustainable development. Academic Press
19. Ellen MacArthur Foundation (2015) Growth within: a circular economy vision for a competitive Europe. Ellen MacArthur Foundation 100:81–92
20. Ellen MacArthur Foundation (2017) Case studies. https://archive.ellenmacarthurfoundation.org/case-studies
21. Ellen MacArthur Foundation, Ide (2017) Business model canva. http://circulardesignguide.com/post/circular-business-model-canvas
22. Ellen MacArthur Foundation, McKinsey (2012) Towards the circular economy: economic and business rationale for an accelerated transition. https://content.ellenmacarthurfoundation.org/m/27265af68f11ef30/original/Towards-thecircular-economy-Vol-1.pdf Accessed 20 March, 2024
23. Ellen MacArthur Foundation, McKinsey, SUN (2015) Growth within: a circular economy vision for a competitive Europe. https://www.content.ellenmacarthurfoundation.org/m/3d4eba36b0311c08/original/Growth-within-A-circular-economy-vision-for-a-competitive-europe.pdf Accessed 12 March, 2024
24. Fischer A, Pascucci S (2017) Institutional incentives in circular economy transition: the case of material use in the Dutch textile industry. J Clean Prod 155:17–32
25. Ghaziani A, Ventresca MJ (2005) Keywords and cultural change: frame analysis of business model public talk, 1975–2000. In: Sociological forum, vol 20. Kluwer Academic Publishers-Plenum Publishers, pp 523–559
26. Gilmore S, Sillince J (2014) Institutional theory and change: the deinstitutionalisation of sports science at Club X. J Organ Chang Manag 27(2):314–330
27. Hahn T, Figge F, Aragón-Correa JA, Sharma S (2017) Advancing research on corporate sustainability: off to pastures new or back to the roots? Bus Soc 56(2):155–185
28. Hahn T, Pinkse J, Preuss L, Figge F (2015) Tensions in corporate sustainability: towards an integrative framework. J Bus Ethics 127:297–316
29. Haigh N, Hoffman AJ (2014) The new heretics: hybrid organizations and the challenges they present to corporate sustainability. Organ Environ 27(3):223–241
30. Hart SL (1995) A natural-resource-based view of the firm. Acad Manag Rev 20(4):986–1014
31. Hart S (2010) Capitalism at the crossroads: next generation business strategies for a post-crisis world. Prentice Hall
32. Hart S (2012) New product development. In: The marketing book. Routledge, pp 346–373

33. Hart SL, Dowell G (2011) Invited editorial: A natural-resource-based view of the firm: fifteen years after. J Manag 37(5):1464–1479
34. Hart SL, Milstein MB (1999) Global sustainability and the creative destruction of industries. MIT Sloan Manag Rev 41(1):23
35. Hasse R, Krücken G (2008) Systems theory, societal contexts, and organizational heterogeneity. In: Greenwood R, Oliver C, Suddaby R, Sahlin K (eds) The SAGE handbook of organizational institutionalism. Sage, pp 539–560
36. Hawken P, Lovins A, Lovins L (2000) Natural capitalism: the next industrial revolution. Earthscan
37. Hodgson GM (1998) The approach of institutional economics. J Econ Lit 36(1):166–192
38. Hoffman AJ (1999) Institutional evolution and change: environmentalism and the US chemical industry. Acad Manag J 42(4):351–371
39. Hoffman AJ, Jennings PD (2015) Institutional theory and the natural environment: research in (and on) the Anthropocene. Organ Environ 28(1):8–31
40. Jones GM (1960) Educators, electrons, and business models: A problem in synthesis. Account Rev 35(4):619–626
41. Lacy P, Keeble J, McNamara R, Rutqvist J, Haglund T, Cui M, Cooper A, Pettersson C, Kevin E, Buddemeier P (2014) Circular advantage: innovative business models and technologies to create value in a world without limits to growth. Accenture
42. Lacy P, Long J, Spindler W, Lacy P, Long J, Spindler W (2020) Introduction: the path to transformation is circular. In: The circular economy handbook: realizing the circular advantage, pp 1–14
43. Lacy P, Rutqvist J (2015) Waste to wealth: the circular economy advantage, vol 91. Palgrave Macmillan, London
44. Lewandowski M (2016) Designing the business models for circular economy—towards the conceptual framework. Sustainability 8(1):43
45. Lifset R, Boons F (2011) Industrial ecology: business management in a material world. Oxford University Press. https://doi.org/10.1093/oxfordhb/9780199584451.003.0017
46. Linder M, Williander M (2017) Circular business model innovation: inherent uncertainties. Bus Strat Environ 26(2):182–196
47. Linnenluecke MK, Griffiths A (2013) Firms and sustainability: mapping the intellectual origins and structure of the corporate sustainability field. Glob Environ Chang 23(1):382–391
48. Lounsbury M, Fairclough S, Lee MP (2011) Institutional approaches to organizations and the natural environment. Oxford University Press. https://doi.org/10.1093/oxfordhb/9780199584451.003.0012
49. Lüdeke-Freund F, Gold S, Bocken NM (2019) A review and typology of circular economy business model patterns. J Ind Ecol 23(1):36–61
50. Maguire S, Hardy C, Lawrence TB (2004) Institutional entrepreneurship in emerging fields: HIV/AIDS treatment advocacy in Canada. Acad Manag J 47(5):657–679
51. Mentink B (2014) Circular business model innovation: a process framework and a tool for business model innovation in a circular economy. http://resolver.tudelft.nl/uuid:c2554c91-8aaf-4fdd-91b7-4ca08e8ea621
52. Meyer J, Rowan B (1977) Institutionalized organizations: formal structures as myth and ceremony. Am J Sociol 83(1):340–363
53. Montiel I, Delgado-Ceballos J (2014) Defining and measuring corporate sustainability: are we there yet? Organ Environ 27(2):113–139
54. Moser F, Jakl T (2015) Chemical leasing—a review of implementation in the past decade. Environ Sci Pollut Res 22:6325–6348
55. North D (1990) Institutions, institutional change and economic performance. Cambridge University Press
56. Oliver C (1991) Strategic responses to institutional processes. Acad Manag Rev 16(1):145–179
57. Osterwalder A, Pigneur Y (2010) Business model generation: a handbook for visionaries, game changers, and challengers, vol 1. John Wiley & Sons

58. Osterwalder A, Pigneur Y, Tucci CL (2005) Clarifying business models: origins, present, and future of the concept. Commun Assoc Inf Syst 16(1):1
59. Planing P (2015) Business model innovation in a circular economy reasons for non-acceptance of circular business models. Open J Bus Model Innov 1(11):1–11
60. Randles S, Laasch O (2016) Theorising the normative business model. Organ Environ 29(1):53–73
61. Salvadó JA, de Castro GM, Verde MD, López JEN (2012) Environmental innovation and firm performance: a natural resource-based view. Springer
62. Scott JT (2017) The sustainable business: a practitioner's guide to achieving long-term profitability and competitiveness. Routledge
63. Scott W (1995) Institutions and organizations. Sage
64. Scott W (2008) Institutions and organizations: ideas and interests, 3rd ed. Sage
65. Sequeira AH (2014) Conceptualisation in research. SSRN Electron J. https://doi.org/10.2139/ssrn.2489284
66. Shrivastava P, Hart S (1995) Creating sustainable corporations. Bus Strateg Environ 4(3):154–165
67. Starik M (1995) Should trees have managerial standing? Toward stakeholder status for non-human nature. J Bus Ethics 14:207–217
68. Starik M, Kanashiro P (2013) Toward a theory of sustainability management: uncovering and integrating the nearly obvious. Organ Environ 26(1):7–30
69. Teece DJ (2010) Business models, business strategy and innovation. Long Range Plan 43(2–3):172–194
70. Tukker A (2004) Eight types of product–service system: eight ways to sustainability? Experiences from SusProNet. Bus Strateg Environ 13(4):246–260
71. Van Ostaeyen J, Van Horenbeek A, Pintelon L, Duflou JR (2013) A refined typology of product–service systems based on functional hierarchy modeling. J Clean Prod 51:261–276
72. Van Renswoude K, Wolde AT, Joustra DJ (2015) Circular business models. Part 1: An introduction to IMSA's circular business model scan. https://groenomstilling.erhvervsstyrelsen.dk/sites/default/files/media/imsa_circular_business_models_-_april_2015_-_part_1.pdf
73. Vatn A (2005) Institutions and the environment. Edward Elgar
74. Webster K (2013) A concise guide to the circular economy. In: EMF (ed) A new dynamic. Effective business in a circular economy, pp 19–28. https://www.ellenmacarthurfoundation.org/a-new-dynamic-effective-business-in-acircular-economy
75. Waddock S (2011) We are all stakeholders of Gaia: a normative perspective on stakeholder thinking. Organ Environ 24(2):192–212
76. World Wildlife Fund (2016) Living planet report. http://Assets.Wwf.Org.Uk/Custom/Lpr2016/ Accessed 20 October, 2023
77. WRAP (2015) Innovative Business Models. Available online: http://www.wrap.org.uk/content/innovative-business-models-1 Accessed 24 April 2024
78. WRAP (2016) Innovative business models. http://Assets.Wwf.Org.Uk/Custom/Lpr2016/
79. Zairul M, Wamelink JW, Gruis V, John L (2015) New industrialised housing model for young starters in Malaysia. The Asia Pacific Network for Housing Research. https://www.semanticscholar.org/paper/New-Industrialised-Housing-Model-forYoung-Starters-M.N-Wamelink/f1f27689b83e98462558806b7b6bca0f04d84382
80. Zott C, Amit R, Massa L (2011) The business model: recent developments and future research. J Manag 37(4):1019–1042

Chapter 5
Gaps in Circular Economy Business Model Theoretical Frameworks

Abstract This chapter highlights the gaps identified in previous conceptual frameworks of the Circular Economy Business Model (CEBM), which were not thoroughly evaluated as comprehensive constructs. These gaps form the basis for the new constructs in the conceptual framework of the current study, namely, materials information management systems (MIMS) and organisational culture. The study draws on contingency theory and dynamic capability theory to address the gap related to materials information management systems. It is established that information technology (IT), which has significantly transformed business management, can enhance the effectiveness and efficiency of materials management. Information, being processed data that adds value to the recipient, supports the need for a computer-based MIMS. Such a system is crucial for scientific inventory control and model implementation. The use of materials identification codes and location codes facilitates quick identification and retrieval of materials from storage, thereby improving operational efficiency. The materials management system also aligns with the circular economy (CE) by improving financial performance. It provides firms with a competitive advantage, leading to improved financial outcomes. MMIS has been found to boost financial success by driving stakeholder relationships, increasing motivation, satisfaction, and loyalty while attracting investors and clients and enhancing the firm's market legitimacy. Several theories were referenced in addressing the organisational culture gap, including social identity theory and change management theory. The Quinn Competing Values Framework (CVF) and Kotter's change model were also considered. Kotter's model emphasises that change is an ongoing, adaptive process to evolving conditions. This perspective on change highlights the need for continuous training and education for employees, beyond job-related skills, to include innovation, technology, methods, material selection, professional conduct, and values. The model advocates for a structured approach to change, where construction organisations develop a clear plan to integrate CE into their business models. Strengthening organisational adaptability increases the likelihood of success. Based on years of research, Kotter's eight-step approach guides organisations in navigating change and improving performance in a dynamic environment. Lastly, for the construction industry to adopt CE, it must address changes in values, mindsets, behaviours, competencies, and capabilities.

© The Author(s), under exclusive license to Springer Nature Switzerland AG 2025
O. Otasowie et al., *Circular Economy Business Model for Construction Organisations*, Green Energy and Technology, https://doi.org/10.1007/978-3-031-88322-4_5

5.1 Introduction

Drawing from the models of Lewandowski [127], Lüdeke-Freund et al. [135], EMF and IDEO [67], as well as Antikainen and Valkokari [14], all of which were grounded in core business model elements and adapted to align with circular economy (CE) principles, Lewandowski's [127] Circular Economy Business Model (CEBM) stands out as the most extensive framework. This model emphasises a closed-loop production process, where materials or products are recovered through a take-back system rather than discarded at the end of their lifecycle. However, concerns arise about the model's practical applicability since it is entirely derived from a literature review and lacks empirical validation. Furthermore, Antikainen and Valkokari's [14] framework has been tested in a real-world scenario involving a Finnish social enterprise. However, its application revealed significant challenges, notably the time-intensive implementation. In addition, research on conceptualising CEBMs remains limited in academic literature. Consequently, industries like construction, the focus of this study, suffer from inadequate research. This gap suggests that existing frameworks fail to address crucial factors influencing CEBMs within the construction sector.

Since frameworks shape how problems are identified and solutions are devised, this section explores two key gaps: materials management systems and organisational culture. Addressing these gaps recognises that the successful implementation of a CEBM depends on robust systems for managing materials and an organisational culture conducive to integrating CE principles into construction operations.

5.2 Gap One: Materials Information Management System

Before concluding how CEBMs could be implemented within the construction industry, examining the underlying reasons for discrepancies between existing CEBM research findings is essential. This includes assessing whether the theoretical frameworks previously proposed for other industries adequately address the construction sector's unique challenges. As highlighted in the literature, CEBMs are typically conceptualised by adapting the traditional Business Model (BM) canvas to incorporate the principles of CE. The objective is to balance the interests of all stakeholders, not just shareholders, while achieving profitability without causing environmental harm. CEBMs aim to bridge the gap between theory and practice, providing a framework for integrating CE into business operations. For instance, the BM canvas explains how companies create, deliver, and capture value [163, 216], forming the foundation for translating CE principles into actionable strategies. One study on CEBMs emphasises integrating components across macro (global trends and drivers), meso (ecosystem and value co-creation), and micro (organisations, clients, and consumers) levels [234]. For example, macro-level factors like evolving regulations can significantly influence business models. Lewandowski [127] focused on adopting closed-loop production processes, where materials or products are recovered at the end of their useful life through systems such as take-back programs. The study identified factors critical for aligning business models with CE principles. It argued that a targeted focus on key BM components is necessary for predicting and achieving successful CEBM outcomes.

However, a recurring limitation in many CEBM studies is an overemphasis on profit-driven variables, often at the expense of addressing fundamental infrastructure needs, such as materials management systems. Lewandowski [127] framework acknowledged the importance of adopting information technology (IT) and data management systems to track materials effectively. Similarly, researchers like Scott [199], Laubscher and Marinelli [125], and Nilsson and Söderberg [155] emphasised that material information management systems are critical for CEBMs, particularly in construction. They argued that these systems should complement the BM canvas, extending beyond profit-focused elements to support lifecycle tracking and management of materials. Peters et al. [168] further argued that a robust material information management system is essential for any CEBM framework in the construction industry, enabling data collection, tracking, and management throughout a material's lifecycle. Hence, there is a need to explore the role of IT and data management systems in enhancing CEBM frameworks for the construction sector.

5.2.1 Materials Management

Efficient materials management plays a vital role in enhancing the productivity of construction projects. According to Bell and Stukhart [20: 15], the key responsibilities of materials management include the following: "material requirement planning and material take-off, vendor evaluation and selection, purchasing, expenditure, shipping, material receiving, warehousing and inventory, and material distribution". The process involves planning and regulation to ensure that materials meet the required quality and quantity specifications, are procured at reasonable prices, and are available when needed. Materials management focuses on coordinating the logistics within a supply chain, encompassing the transportation and storage of raw materials and managing inventories of items in progress and finished goods from their origin to the point of use. This management begins during the project's planning phase and continues until its completion. Effective materials management is especially critical for construction projects, where sites typically handle significant volumes of heavy and expensive supplies. Therefore, implementing a robust system to oversee materials management is essential.

5.2.2 Materials Management in the Construction Industry

The construction sector represents the largest segment of the global economy, and projects that meet their timelines, budgets, design specifications, and stakeholder expectations are typically deemed successful [154]. Construction encompasses the physical assembly of structures and the organisation and deployment of tools, materials, management, supervision, and equipment necessary for completion. Executing a construction project demands collaboration among multiple stakeholders, including clients or owners, architects, engineers, contractors, suppliers, and vendors. It involves the intricate and often demanding task of transforming resources, such as machinery, finances, materials, and labour, into substantial, immovable assets. Hence, material management is a critical factor influencing the successful delivery of construction projects. As projects increase in scale and complexity, managing materials becomes more challenging,thus, practical tools and strategies are essential to ensure timely material supply, accurate inventory tracking, schedule adherence, and waste reduction. Larger and more sophisticated projects require advanced approaches to material management, given the heightened coordination demands and the project's significance. Also, proper planning is crucial to address these challenges, mainly when dealing with complex projects. Nevertheless, the adoption of advanced information and communication technologies (ICT) has the potential to revolutionise material management processes. Emerging technologies such as wireless communication and tagging systems could significantly enhance future material handling and management.

Furthermore, poor material management remains a primary cause of inefficiencies and disruptions in construction [182]. Inefficient handling and management of materials can severely impact project costs, schedules, and quality. Materials typically account for 50–60% of total project costs [32], with some estimates suggesting material management expenses could constitute 30–80% of the total budget, depending on the type of construction [147]. Given its substantial impact on project outcomes, effective material management is vital to minimise delays, maintain quality, and control costs. Delayed material delivery has been identified as a significant contributor to project time overruns [58]. Other factors influencing delays include inadequate material planning, transportation challenges, material shortages, and insufficient communication systems. These issues can hinder site operations and affect overall project performance [112]. Hence, seamless integration between the design phase and material utilisation is essential to mitigate such delays. Implementing robust ICT systems and efficient management practices can facilitate this transition and improve project outcomes.

5.2.3 Processes of Materials Management in the Construction Industry

Materials management procedures encompass various aspects of construction projects, including logistics, inventory and waste management, planning, procurement, handling, and oversight of materials. Driving a favourable environment for materials management on construction sites ensures effective handling practices. To support materials management, the following strategies are highlighted: managing the planning, acquisition, transportation, handling, storage, and disposal of materials.

5.2.4 Planning

Planning construction techniques involves understanding the project's requirements, selecting the most suitable methods, and executing them cost-effectively to meet client expectations [105]. A well-defined plan is essential for achieving the objectives of any task, particularly in engineering, construction, and related activities. Effective material planning guides subsequent operations and determines the project timeline. It includes tracking and maintaining records of all materials used in a facility, setting target inventory levels, and ensuring the regular supply of resources. Proper record management facilitates the smooth flow of materials on-site, reducing the risk of shortages or undelivered items.

Stukhart [208] emphasises that material planning is a foundation for subsequent tasks and can significantly influence the project schedule. This process outlines inventory targets, supply schedules, and record-keeping requirements [167]. However, inadequate integration of material planning with the construction schedule often makes materials unavailable when needed [112]. An efficient materials management strategy improves productivity and profitability and ensures timely project completion. Effective planning for material accessibility and routing within a construction site directly impacts the success of a project. Enhanced material planning reduces activity durations and project delays and improves service quality, which is essential for boosting organisational productivity and profitability.

5.2.5 Procurement

Procurement encompasses various activities, including acquiring tools, supplies, labour, and services for executing and implementing a project. According to Eriksson and Lind [69], procurement operates within an inter-organisational, project-based framework, where clients and suppliers collaborate to generate sustainable value. In materials management, procurement aims to ensure the availability of high-quality materials at the right time, location, and cost. This involves organising material purchases, informing suppliers of delivery schedules, and tracking deliveries to confirm deadlines. Canter [35] posits that inefficient procurement processes can lead to excessive material costs and over-ordering.

The procurement of materials begins with defining the project's requirements. This is followed by selecting suitable suppliers or subcontractors and concludes with the delivery of materials to the specified location. Many construction companies face challenges in sourcing materials from reliable suppliers at reasonable prices with timely delivery. A control plan is necessary to achieve the desired outcomes throughout the procurement process. All purchase requests and quotations should follow an authorised requisitioning procedure, typically managed by the project manager. The project manager's role includes ensuring that materials are procured in alignment with project specifications, timelines, and quality standards.

5.2.6 Logistics

Logistics focuses on mobility and involves planning, executing, and managing the movement and storage of materials, ranging from raw resources to finished products, to meet client needs [208]. Materials often vary in size and weight in construction, necessitating careful handling throughout the supply chain. This requires efficient transportation of materials from suppliers to construction sites, both in factories and at project locations [79]. Effective logistics strategies in construction projects emphasise improved communication and collaboration among project stakeholders during the design and construction phases, particularly in managing the flow of materials [223]. However, the process of material flow management faces challenges such as delays caused by just-in-time ordering and material wastage during storage, handling, and transportation when excessive quantities are procured without considering on-site requirements.

Material routing significantly influences the duration and cost of construction projects [209]. Therefore, accurate forecasting of material movement and meticulous planning of site access and circulation are essential to achieving efficient material management within the logistics framework.

5.2.7 Handling

Tompkins et al. [225] highlight that effective material handling relies on appropriate techniques to ensure materials are distributed in the correct volume, sequence, timing, condition, location, and cost. Construction projects involve proper management, handling, and storage of materials. Efficient material handling facilitates the movement and positioning of materials systematically and contributes to the design of streamlined systems. Material handling encompasses all aspects of placing and transporting items, making it critical to ensuring efficiency and cost-effectiveness.

Material handling is often regarded as an expensive operation, requiring strategic decision-making. According to Temiz and Calis [221], the choice of machinery plays a pivotal role, as it can lead to optimised labour use, enhanced production processes, greater system adaptability, and increased output. Quality considerations must also be integrated into material handling systems, given the frequent movement of materials and its potential impact on overall project quality.

Furthermore, effective material handling requires deliberate planning and adequate financial resources. Hampton et al. [93] posit that material handling costs can range from 10 to 80% of the total expenses in construction projects, depending on the type of facility. Hence, ensuring that materials are delivered to the site without damage or loss is a fundamental objective of proper handling. However, inadequate facilities for handling and unloading materials often lead to significant waste [72]. Thus, safe handling practices on-site are essential to minimise waste, boost overall system productivity, and ensure the smooth flow of operations.

5.2.8 *Stock and Control of Waste*

The European Construction Institute's Total Productivity Management study [71] highlights the critical role of material delivery in construction productivity, emphasising the need for a quality control system for early monitoring and oversight. Proper stock management is essential, particularly for frequently delivering building materials. An adequate stock control system ensures the availability of all necessary items, including raw and processed materials, assembly components, general and consumable stores, and spare parts whenever required [149]. However, construction projects can generate significant amounts of waste, with material waste identified as a major issue in the sector [149, 230]. It is reported that construction waste accounts for around 29% of total waste in the USA, over 50% in the UK, and between 20 and 30% in Australia. This is mainly due to the variety of materials used during construction. Waste can occur at any project stage, from planning and design to operation and construction [38]. Therefore, waste can be minimised by focusing on material reuse and reduction during the design and construction phases [72, 189, 211, 214]. In addition, careful material storage on-site is vital to prevent waste, material damage, and other losses, ensuring that construction work proceeds smoothly. Effective material management reduces profit loss from waste, theft, damage, and stock shortages. Consequently, a strong management system should begin at the start of the building process to ensure optimal outcomes.

5.2.9 *Materials Management Issues*

Improper materials management in construction projects arises from a variety of factors. According to Nasir [150], challenges such as transportation issues, waste, excessive paperwork, improper handling of materials on-site, misuse of specifications, inadequate supply of resources, and a lack of proper planning all contribute to the negative impact on material management. Delays in material management can lead to shortages of materials on construction sites [1]. In addition, material management is often disrupted by the late arrival of essential materials. Kasim [113] also highlighted several common challenges related to material management, which include:

(i) Receiving items earlier than required can lead to increased inventory costs and a higher risk of material quality deterioration.
(ii) Delays in material deliveries negatively affect production output.
(iii) Delayed initiation of projects due to incomplete or unavailable design documents.
(iv) Additional design changes or revisions.
(v) Risk of items being damaged or lost.
(vi) Choosing the appropriate contract type for procuring a single material.
(vii) Establishing criteria for assessing suppliers' performance.
(viii) Managing stockpiles and inventory efficiently.
(ix) Handling excess or surplus materials effectively.

Traditional construction methods rely on paper-based processes, leading to excessive paperwork and poor material management in building projects [150]. This approach can be inefficient and prone to errors, especially when managing the supply chain's exchange and documentation of material components. However, adopting Information and Communication Technologies (ICT) can enhance the efficiency and speed of building operations. ICT advancements have transformed outdated practices, improving material management systems. Creating new software tailored specifically for the construction industry has further supported more effective material management. ICT-enabled solutions can address material management challenges, such as using innovative technologies to streamline material orders, requests, bidding, and waste management, ultimately improving supply chain efficiency.

5.2.10 Information and Communication Technologies in Materials Management

Advancements in ICT have significantly accelerated the global transfer of information compared to traditional methods in the construction industry. Perera et al. [169] define ICT as encompassing software, electronic devices (such as computers), and tools like telephones, fax machines, and photocopiers that facilitate information processing, transmission, storage, and display. Historically, key documents such as specifications and drawings were primarily shared in paper format. However, while industries like publishing, manufacturing, and advertising have rapidly embraced ICT to enhance their global operations, the construction sector is increasingly leveraging technology to support various building activities. Griffith and Stephenson [87] highlight that construction firms investing in ICT innovation have significant growth opportunities. Incorporating ICT into materials management can streamline and optimise on-site materials control. Typical applications include using software such as Lotus 1–2–3 and MS Excel for cost estimation [169, 210]. In addition, ICT facilitates e-commerce practices, including electronic payments, invoicing, and receipts, often conducted through the Internet and email platforms [169]. By transitioning from conventional methods to more advanced and innovative technologies, contractors and suppliers can enhance the efficiency of procurement and ordering processes. Furthermore, there is a growing opportunity to adopt computer-based solutions to improve materials management on construction sites [159]. Various applications developed by researchers have demonstrated the potential of ICT in this area.

5.2.11 Building Information Modelling (BIM)

BIM is a concept that combines spatial and ICTs and is widely used in the architecture, engineering, and construction (AEC) industry. It can be systematically integrated with technologies for data identification and collection [151]. BIM minimises errors, redundancy, and unexpected design changes by enhancing the quality and precision of design and construction processes. It supports better coordination and communication in collaborative projects, enables simulation and analysis, facilitates lifecycle information evaluation, and promotes sustainable design throughout a project's lifecycle [132]. Functioning as a digital repository, BIM simplifies the management of project-related information. Various studies have explored the integration of BIM with other management approaches to reduce waste. For instance, Won et al. [244] utilised BIM to develop a material database for building demolition projects. Similarly, Hamidi et al. [92] proposed using BIM to establish a system to manage material waste after demolition. Cheng and Ma [41] used BIM to estimate demolition waste and associated disposal costs and optimise vehicle scheduling for waste collection. BIM allows stakeholders and users to share material-related information and implement knowledge-based strategies effectively through these capabilities.

5.2.12 Enterprise Resource Planning (ERP)

Computer systems' capabilities and associated software applications have advanced considerably. Tools such as ERP systems and technologies that enable businesses to connect electronically (e.g., via the Internet) now allow companies to exchange vast amounts of information quickly and efficiently. Rapid access to information has become critical for many organisations to maintain market competitiveness. Statistics from the Central Statistical Office in Poland [90] indicate that 2019 over 90% of large enterprises and more than 61% of medium-sized companies in Poland utilised

ERP-class software. ERP systems are primarily designed to record and manage the economic transactions occurring within an organisation. They serve as a centralised platform for collecting, organising, maintaining, and sharing data across all functional areas of a business.

In the manufacturing sector, one of the key modules of an ERP system is material requirements planning (MRP), which operates within a closed-loop control framework. While MRP focuses on detailed planning of material needs, the ERP system ensures seamless integration across all departments of an organisation. Unlike traditional inventory control methods, the MRP system relies on the dependent demand principle, meaning inventory and production capacity are driven by orders for finished goods. The material demand algorithm requires the following to operate effectively [90]:

(i) A comprehensive bill of materials.
(ii) Details on inventory levels, conditions, manufacturing cycles, and the supplier's pricing information.
(iii) Production plans outlining the manufacturing phases, specifying completion timelines, production quantities, and readiness status; and
(iv) Data on production routes, standard time requirements, and any restrictions related to production orders.

5.2.13 Construction Materials Planning System (CMPS)

The primary objective of managing construction materials is to ensure that the correct materials are available in the right quantities and at the appropriate times to meet the demands of planned construction activities. Consequently, CMPS must identify the necessary components to meet the Master Construction Schedule (MCS) and calculate the timeframes within which these components should be available, considering the lead times required for their procurement [245]. Tasks such as determining what to order, the quantity to order, the timing for orders, and scheduling deliveries fall under the responsibilities of a CMPS. To maintain an up-to-date MCS, a CMPS must also be capable of reprioritising tasks. The availability and demand for materials and components can fluctuate significantly due to various factors. Changes or additions to project specifications by clients, delays in material deliveries, shortages, scrap generation, or equipment failures are common challenges. In such cases, the CMPS collaborates with the materials controller to manage tasks like adding, expediting, modifying, or cancelling orders to align with current project priorities.

While Material Planning Systems (MPS) techniques are not new, their application in small-scale construction projects was historically impractical due to economic constraints. Over the past two decades, advancements in computer technology and a significant reduction in hardware costs have made such systems more accessible. However, implementing an MPS in construction projects requires substantial data storage, retrieval, and processing capabilities, which are now achievable with modern computer systems.

5.2.14 Construction Materials Exchange (COME)

COME is an e-commerce platform facilitating seamless electronic communication among agents, buyers, and suppliers. It enables direct interaction between vendors and clients, driving negotiations and promoting competitive business practices. Buyers can efficiently search for, compare, and directly contact suppliers' products through the platform. In addition, buyers may request agents to assist with specific tasks required to complete transactions when needed. The COME platform features an extensive e-catalogue that categorises over 2000 types of building materials into 17 hierarchical trees, each with up to four levels of detail. This structure allows users to start their product searches broadly and narrow down to more specific information as required [117]. The system also provides information on relevant materials in the region, including associated transportation costs for delivery from suppliers to clients. Buyers are empowered to make material requests, review and accept supplier bids, select agents to manage orders and purchase supplier contact information. They must decide whether to engage an agent to complete their order or contact the suppliers by acquiring their details. Suppliers can use the COME system to update the e-catalogue with new products and business details, analyse market trends related to material prices and buyer preferences, and submit bids in response to client requests. Agents, on the other hand, can manage client orders and place bids to fulfil buyer requirements.

5.2.15 Bar-Code System

Bar-coding technology is a widely adopted and well-established tool for tracking and identifying materials across various industries worldwide. It is more commonly utilised than automated identification (auto-ID) technology. In the construction industry, barcoding has been used to manage materials, equipment, and tools since 1987 [39]. The implementation of this technology enhances the efficiency of material management processes. Inventory managers and supervisors can instantly exchange real-time information regarding item quantities using barcoding. In addition, barcoding offers several key features, including the following:

(i) Monitoring construction site materials in real-time
(ii) Automatically logging historical data related to building materials used in the project;
(iii) Continuously tracking materials utilised by project teams and
(iv) Enabling automatic data transfers, in real-time, through intranet or internet connections to the headquarters.

Hence, Finkenzeller [73] describes a barcoding system as a binary coding method consisting of parallel narrow and broad bars separated by gaps. This system allows data to be represented in both numeric and alphanumeric formats. Key components of the bar-coding system include the barcoding labels, a printer to produce the codes, and a laser scanner, which is used to read and interpret the information encoded on the labels.

5.2.16 Radio Frequency Identification (RFID)

RFID is a technology employed in material management to identify objects or individuals using radio waves. Hunt et al. [104] describe RFID as a wireless communication technology designed to identify tagged items or people uniquely. RFID operates as a data carrier within information management systems, facilitating data exchange between a host system and the RFID tag [73]. The adoption of RFID technology has seen significant growth across various industries, including distribution, manufacturing, and retail. Valero et al. [233] highlight its potential benefits, such as enhancing quality, productivity, and safety while optimising project timelines and reducing costs associated with materials and labour. Wood and Alvarez [246] noted that RFID monitors and manages tagged materials in the construction sector as they navigate supply chains and building sites. The study also provides specific examples of its application in materials management, which include:

(i) Tracking Material Inventory: The implementation of RFID technology enables real-time, automated, and accurate tracking of specialised materials across the entire supply chain, from manufacturing and transit to delivery, reception, storage, and installation. With RFID, materials can be automatically identified as they are transported and received, ensuring that data is seamlessly updated and transmitted to materials management systems without requiring manual entry, reducing the risk of human error.

(ii) Streamlined Materials Tracking and Expediting: Adopting an RFID system offers numerous benefits over traditional manual tracking, including providing timely, precise information for field planners. This allows for better scheduling, optimising critical jobs, and quickly identifying available resources for crafts that may be underutilised temporarily.

(iii) Accurate Inventory and Material Status Records: To avoid missing, misplaced, or undelivered materials that could delay project timelines, RFID technology provides accurate information about the shipment, delivery, and current status of materials. For instance, a handheld scanner can verify material deliveries, ensuring precise and up-to-date inventory management.

5.2.17 Theoretical Underpinnings of Materials Information Management System

To comprehend and address the material management system gap within the CEBM, the study will draw on the Technology Acceptance Model, Contingency Theory, and Dynamic Capability Theory.

5.2.18 Technology Acceptance Model (TAM) Theory

Ajzen and Fishbein's [13] Theory of Reasoned Action (TRA) has been expanded in several notable ways, with the Technology Acceptance Model (TAM) being one of the most influential adaptations. TAM was developed by Fred Davis and Richard Bagozzi [15] and is notably recognised for its focus on two key factors: perceived usefulness (PU) and perceived ease of use (PEOU), which replace many of the psychological state elements in TRA. TRA and TAM emphasise that when individuals decide to engage in a specific action, they typically do so without significant barriers. However, potential constraints, such as limited resources, time restrictions, regulatory limits, or lack of awareness, may impede their ability to act [15]. Perceived usefulness (PU) refers to the degree to which an individual believes using a particular system will improve performance. Perceived ease of use (PEOU) reflects the belief that using the system will require minimal effort. Venkatesh and Davis [237] defined value as the belief that a particular system would enhance one's performance, while usability was described as the ease with which the system can be used [138]. However, one criticism of TAM is its neglect of external factors that might hinder technology adoption, such as environmental or organisational barriers [12].

Organisations are increasingly adopting information management systems due to the many advantages they offer, including enhanced data management (e.g., spend analysis), seamless integration with other internal systems, faster processing times, improved tracking, fewer errors, and more informed purchasing decisions [178, 224]. Organisations must ensure that users use these platforms to continue benefiting from these systems. Among the various models used to predict the adoption and success of information systems, TAM remains one of the most widely applied studies on organisational user behaviour and system usage success [5, 49].

5.2.19 Contingency Theory

Contingency theory is a framework for understanding organisational behaviour that emphasises the influence of external factors such as technology, culture, and the environment on the design and functioning of organisations. It is founded on the premise that there is no universally optimal organisational structure; an organisation's success depends on the alignment between its size, structure, information systems, and technology, particularly in dynamic environments. The development of contingency theory drew heavily from sociological functionalist theories of organisational structure [42, 181, 247]. These theorists highlighted that organisational size, task environment characteristics, and technological factors significantly shape organisational design.

Research into contingency theory suggests that organisational effectiveness arises from the alignment, or "fit", between various internal and external elements. Van de Ven and Drazin [235] introduced the concept of fit through selection, interaction, and systems. The selection approach posits that an organisation's design should align with its contextual environment to ensure survival and functionality. This perspective, common in early contingency studies, explored the relationship between context and design but did not address organisational performance. For instance, Dewar and Hage [57] examined task and technology in two dimensions, while Freeman [76] investigated technology as a contingent factor. Marsh and Manari [139] demonstrated significant connections between technological attributes and organisational structure. Another interpretation of fit emphasises its role in enhancing performance through the interplay between structure and situational factors. Khandwalla [115] and Van de Ven and Ferry [236] explored this perspective, with the former highlighting that successful organisations exhibit stronger correlations among technology, vertical integration, authority delegation, and sophisticated control systems than less successful ones.

Contingency theory views organisations as open systems that continuously interact with and adapt to their environments [37, 156]. For example, Silva et al. [202] found that using technology in organisational information management can help address challenges posed by change and environmental misalignment. Technological integration supports business growth and innovation, enhancing system efficiency [172]. Chen and Song [40] also identified its role in driving business model innovation and improving competitiveness. Guo et al. [89] further demonstrate that IT management systems enable organisations to align decisions and structures with internal and external demands. Moreover, analysing environmental data can help organisations respond effectively to changing conditions, improving decision-making and enabling adjustments in production capacity to meet market needs [40, 108].

5.2.20 Dynamic Capability Theory

Dynamic capability theory pertains to an organisation's capacity to integrate, evolve, and refine internal and external capabilities in response to a dynamic and ever-changing environment [82, 220]. Initially proposed by Teece and Pisano [219] and later expanded by Teece et al. [220], this framework was introduced to address limitations within the resource-based view theory. Dynamic capability is defined as "the firm's ability to integrate, build, and reconfigure internal and external competencies to address rapidly changing environments" [220: 516]. These strategic processes are critical for organisations operating in competitive markets, enabling them to reallocate resources and innovate to create new value.

Implementing IT data management systems in businesses facilitates the collection of environmental data and the rapid exchanging of real-time, critical information among stakeholders [249]. This enhances decision-making, which relies on informed insights and contributes to developing dynamic data and information processing capabilities [62]. According to Zahra et al. [255], the adaptability of a firm's capabilities significantly impacts its strategic decision-making, often driving innovation. Advanced data storage and processing capabilities allow organisations to make decisions in real-time while also enabling the analysis of data trends over time to better interact with their surroundings [70, 103]. From the perspective of dynamic capability theory, information management systems are seen as essential tools that provide a competitive advantage in volatile and unpredictable environments. In IT data management systems, the term "materials management" is a comprehensive concept aimed at optimising material productivity. Various definitions of this concept have been explored [239] and they include:

(i) "Materials management is managing the flow of materials into, through, and out of the system."
(ii) "Materials management is to ensure adequate supply of materials to meet the anticipated pattern of demand in a system."
(iii) "Materials management pertains to decision making concerning materials to ensure right quantity of materials of right quality at right time at right place in right price to get most out of every dollar invested in them."
(iv) "Materials management is concerned with design, specification, procurement, transportation, inspection, storage, retrieval, use, disposal, and accounting of materials to maximise return on investment in materials."

Materials management is a critical aspect of an organisation's operations. It constitutes a substantial portion of the total cost of goods and services and can often represent more than half of a business's operational expenses. Consequently, managing materials efficiently and effectively becomes pivotal for cost reduction and value creation. For instance, in industries such as cotton textiles, silk yarn, rubber, sugar, automotive, and electrical products, materials may account for 60–70% of total expenditures. In contrast, sectors like pharmaceuticals, nonferrous metals, and construction may see material costs range between 50 and 60%. Considering this significance, ensuring the availability of the right materials in appropriate quantities and at the required times is essential for meeting production targets, achieving service goals, and completing projects on time. Shortages can be more than just frustrating; they can be costly or even catastrophic. Efficient and effective material flow management, from procurement to utilisation and eventual disposal, is imperative, especially in a globally competitive environment where maintaining customer satisfaction and value creation is essential to sustaining profit margins.

Waste management is another crucial aspect of materials management. The effectiveness of a materials management strategy can often be gauged by the amount of waste produced across the supply chain [239]. Effective management minimises waste generation, aiming for zero waste production through strategies aligned with reducing, reusing, and recycling principles. Waste may include surplus materials, discards, or scrap, and an effective system ensures optimal utilisation of resources, thereby maximising the value derived from material expenditures [239]. Hence, integrating information technology (IT) has significantly enhanced the efficiency and effectiveness of materials management. It enables the development of systems that provide scientific inventory control and optimise resource utilisation. According to Vrat [239], processing information into actionable insights is invaluable for improving decision-making. Implementing computer-based materials management systems equipped with materials identification and location codes streamlines processes, making it easier to track, retrieve, and use materials efficiently. These systems improve operational performance and contribute to achieving broader organisational goals.

Information management systems (IMSs) play a vital role in businesses, supporting various operational levels, from everyday transactions to long-term strategic initiatives, within an increasingly digitised environment. These systems align a company's strategy with its operations, ensuring collaboration across multiple divisions and administrative layers to achieve shared objectives. A key focus of IMSs is to provide resource availability and responsible management to maximise stakeholder satisfaction and meet sustainability goals [8, 9]. As discussed in Chap. 3, the stakeholder theory's sequential approach emphasises the need for businesses to devise strategies that successfully integrate circular principles into their business models (BMs). The descriptive perspective of this theory underscores the importance of IMSs in integrating circularity into business operations and strategic objectives, driving a more equitable and prosperous world where cultural and environmental achievements are preserved for future generations [60, 77]. IMSs act as comprehensive tools that help organisations manage the multifaceted nature of computer science (CS) and navigate conflicting areas of interest. By encouraging sustainability, IMSs enhance businesses' ability to comply with regulations, respond to competitive pressures, and address stakeholder expectations [8, 180, 252, 253].

IMSs are particularly valuable for advancing the CE within businesses, improving financial performance, and fostering a competitive edge that drives superior financial outcomes [2, 9]. These systems strengthen stakeholder relationships, boosting motivation, satisfaction, and loyalty [77, 84]. Furthermore, IMSs attract new investors and customers by improving market acceptability and contributing to financial performance through environmentally conscious production, cost reductions, and risk mitigation. Companies excelling in CE often utilise IMSs, reflecting their effectiveness in embedding circularity into BMs. From a stakeholder theory perspective, IMSs provide practical solutions for successfully implementing CE strategies [8, 170].

IMSs support sustainability across social, environmental, economic, and workplace dimensions [183]. They facilitate adopting sustainable practices and preserving the ecological balance in industries such as real estate and construction by reducing costs, resource depletion, and energy consumption [180, 252, 253]. In addition, IMSs enhance organisational capabilities and promote renewable energy use, driving innovation and operational efficiency with positive environmental impacts [8, 251].

Considering these benefits, incorporating IMSs into the framework of CEBMs is crucial, as explored in this chapter.

5.3 Gap Two: Organisational Culture

Research into CEBM is essential to determine how construction organisations can effectively integrate CE principles into their operations. To understand this integration comprehensively, it is crucial to evaluate CEBM from the perspectives of various organisations within the sector. This approach allows for a holistic assessment of how their business models function. Understanding and integrating CE into these models also necessitates a thorough examination of the organisational culture, as cultural factors significantly influence both employee behaviour and the implementation of business strategies. Each organisation possesses a unique culture shaped by its specific beliefs, values, rituals, artefacts, symbols, conventions, and guidelines that influence employee conduct. Contemporary trends like globalisation and digitisation often require organisations to adapt their cultures to remain competitive. However, these changes may encounter explicit or implicit resistance from employees, potentially hindering their success, particularly if they conflict with the organisation's established cultural norms [146].

Organisational culture is defined by shared values and beliefs that govern behaviours and create a sense of belonging among members [68]. A strong culture is essential for the success of an organisation, as it provides guiding principles and drives collaboration to address environmental challenges [231]. Organisational culture aims to implement norms that facilitate goal achievement through collective efforts [3]. Founders or leaders often shape this culture, which evolves as the organisation adapts to internal and external changes [158].

Robert emphasised that organisational culture represents a set of shared standards and values, and active employee engagement in organisational activities reflects the strength of this culture [152]. A strong organisational culture is critical in the construction sector, where it plays a crucial role in driving pride in corporate initiatives and maintaining consistency in principles and practices [212]. The stability of this culture relies on the dedication of employees, who reinforce and uphold core values, creating a foundation for the organisation's overall effectiveness and sustainability.

Furthermore, Grieves [86] highlighted the potential of organisational development to promote values and ideals. In contrast, Hofstede [99] emphasised the significant role of culture in shaping individuals' thoughts and actions, underscoring the importance of understanding an organisation's culture. Deal and Kennedy [52] argued for integrating organisational development and culture to enhance employee efficiency and effectiveness. Martins and Terblanche [141] offered two perspectives on organisational culture: its functional role and influence on organisational processes. According to Martins [140], organisational culture serves two primary purposes. First, it drives a sense of belonging and loyalty among employees. Second, it helps create a competitive advantage by ensuring new and existing employees are familiar with acceptable behaviours and the organisation's social norms. Furnham and Gunter [78] noted that corporate culture provides a shared system of meaning, facilitating communication and understanding among team members. However, a poorly functioning culture can significantly hinder an organisation's effectiveness. Organisations use various tools and practices, such as strategic direction, clear objectives, effective communication, collaboration, decision-making frameworks, and interpersonal interactions, to shape and modify behaviour. These tools indirectly influence behaviour through their connection to organisational culture [141].

Considering the profound impact of culture on organisational operations, it should not be underestimated in the pursuit of efficiency [197]. Morgan [145] described organisations as expressions of human nature, advocating for a people-centric approach to their development. Campbell et al. [34] observed that culture affects many aspects of an organisation, such as employee motivation, morale, productivity, work quality, creativity, and attitudes. Organisational culture can also be a strategic resource to foster growth and success. First, organisational culture can significantly enhance business performance, offering a competitive edge over rivals [30]. Companies like Hewlett-Packard and IBM have leveraged their unique cultures to outperform competitors. Secondly, corporate culture can act as a mechanism for managerial control. Managers can shape and direct employee behaviour through shared values, rituals, symbols, and narratives. This form of control can be cost-effective long-term and promote loyalty to the organisation and its goals.

5.3.1 Concept of Culture

One of the most notable contributions of social and cultural anthropology to modern thought is the concept of culture. When examining this idea, two dominant interpretations emerge, each defined by its role in explanation. The first draws from Edward Tylor's definition, which describes culture as a complex system encompassing "knowledge, belief, art, morals, law, custom, and any other capabilities and habits acquired by man as a member of society" [229: 1]. In this view, culture is a compilation of characteristics requiring explanation, rather than a theoretical construct with explanatory power [187: 390]. The second interpretation is influenced by Émile Durkheim's understanding of social truths, which emphasises collective patterns that transcend individual perspectives. Although Durkheim focused on society in his work, his concept functions similarly to culture as an explanatory framework. For him, collective patterns represent distinct cultural expressions capable of explaining individual behaviour. Consequently, cultures become explanatory entities through these shared patterns.

Over the last century, anthropologists have refined what is now considered the traditional understanding of culture [187]. This perspective conceptualises cultures as cohesive, systematically shared constructs. Figures such as Franz Boas, Bronisław Malinowski, and Alfred Kroeber significantly shaped this approach, influencing other fields like linguistics, sociology, and philosophy. Unlike Tylor and other evolutionists, Boas argued against viewing culture as a collection of arbitrary, isolated traits [186]. Moreover, he rejected evolutionary theories of cultural development, emphasising that culture holds a distinct metaphysical status and should be regarded as a unique phenomenon [26, 133]. While Boas [26] maintained that culture comprises deeply ingrained characteristics, he advanced the understanding of culture by exploring its interactions with its elements and the physical environment. This perspective represented a significant step towards a more holistic comprehension of cultural phenomena.

Furthermore, the influence of philosophy on anthropology is notable. Dilthey [59] distinguished between civilisation and the study associated with the realm of reason and nature, which is linked to the domain of causality. According to the study's philosophical perspective, biology and culture possess fundamentally different ontological statuses. This distinction has significantly shaped ethnography's theoretical foundations and practical applications. Within this framework, society is understood as existing for multiple reasons, necessitating that social phenomena be interpreted within their social contexts rather than explained through natural scientific causes. Consequently, Boas [26] approached cultures as rational constructs. By doing so, their theoretical lens blurred the boundaries between scientific and social sciences, effectively challenging the dichotomy between the two fields. This divergence marked the emergence of a "theoretical apartheid" in cultural anthropology, where researchers from the humanities distanced themselves from those in the natural sciences, perceiving the two disciplines as fundamentally distinct in their ontological assumptions [25].

In addition, the concept of configurationalism, advanced by scholars like Kroeber, Mead, and Benedict, pushed the metaphysical understanding of culture to new heights [95, 227]. This approach proposed that culture is shaped according to enduring patterns, separating it from individuals and presenting it as an independent ontological entity akin to natural laws in physics. However, this perspective faced criticism, particularly its failure to explain the interplay between individual psychology and cultural structures. Although Kroeber's configurations avoided strict determinism, they implied a degree of stability in cultural elements that risked creating an impression of determinism by suggesting causal links between individual behaviours and fixed cultural traits [95, 227]. However, two primary approaches emerged after World War II for understanding culture. The first group of scholars, materialists [95] or realists [22, 23], viewed culture as inseparable from human populations, emphasising it as a collection of characteristics or a mode of social interaction [22: 449, 187: 393]. The second group, idealists, defined culture through laws, customs, and beliefs [22, 23, 95]. These idealists were influenced by diverse theoretical frameworks, including structuralism, ethnoscience, functionalism, structural-functionalism, and symbolic anthropology [187]. They conceptualised culture as a unified and coherent system, culminating in the modern understanding of culture as a collection of homogeneous entities, each characterised by consistent beliefs, values, knowledge, or behavioural patterns [187].

Clifford Geertz's "thick description" concept represents a cornerstone of the traditional understanding of culture [81]. The study explored the connection between meaning and behaviour, asserting that meaning is conveyed through actions and becomes intelligible only when these behaviours are interpreted within their societal context. The study proposed that culture could be seen as a framework of rules, shaping how individuals interact. This framework enables explanations like "Because of culture," though such answers often oversimplify complex dynamics. However, Geertz's concept of culture and the framework faced significant criticism. First, for culture to function effectively as a theoretical construct, its boundaries must be clearly defined. Critics argued that researchers often arbitrarily determined these boundaries [47]. Second, anthropologists' cultural usage frequently implied a homogenous set of customs and norms, whereas ethnographic evidence showed that such customs varied even within small communities. Third, the portrayal of culture as a unified set of rules suggested individuals were merely passive followers of these norms, an idea challenged by critics who emphasised human agency [176, 187]. Lastly, the tendency to view cultural norms as rigid structures limits understanding diverse behaviours and individual negotiations within societies [190]. This perspective often overlooked the complexity of human agency and the dynamic interplay between individual and collective practices [27, 28, 254]. For culture to remain a meaningful explanatory concept, it must reflect its real-world effects on individuals. Nevertheless, culture can only be understood indirectly through its manifestations, necessitating hypotheses that interpret these expressions. Some argue that culture, as a theoretical construct, functions more as a rhetorical device than as a concrete explanatory tool. This has led anthropologists to rethink the concept of culture, acknowledging its limitations while striving to develop alternative theoretical frameworks, such as neo-Boasian approaches [19], practice theory [27], and the ontological turn [102]. These frameworks allow scholars to focus on specific areas, such as rituals, local ontologies, or kinship systems, while sidestepping the pitfalls of traditional cultural theory.

In contemporary evolutionary studies, three primary perspectives on cultural development have emerged. The first focuses on "pure" cultural evolution, likening cultural units to genes that replicate and spread. This perspective includes the concept of "memes," popularised by Richard Dawkins and Daniel Dennett [24, 50, 56, 65, 128, 129]. The second perspective adopts a historical approach, defining culture as the heritable transmission of non-genetic traits that influence reproductive success. A prominent example of this is the increased prevalence of certain genetic disorders in the Ashkenazi Jewish population, shaped by cultural practices [88, 222]. The third perspective encompasses theories within cultural evolutionary science (CES), emphasising how small, incremental changes in individual behaviours can collectively lead to large-scale cultural patterns. This includes dual-inheritance theory [98, 143, 184], cultural epidemiology [10, 29, 205], and gene-culture coevolution [63, 123, 124]. These theories propose that biological and cultural evolution are interdependent, with mutual influences shaping human development. A well-documented example is the coevolution of dairy farming practices and the emergence of adult lactose tolerance [63, 97, 185]. Through these evolving perspectives, the concept of culture continues to adapt, offering nuanced ways to study the intricate relationship between human behaviour, society, and biology.

5.3.2 Organisation Culture Definition

Since the 1980s, Japanese firms have demonstrated the significance of strong, cohesive organisational cultures in driving productivity and competitiveness. Their remarkable success during this period, particularly in surpassing Western firms, highlighted the relationship between organisational culture and performance [80]. Unlike Western businesses, which often emphasise individual functionality, Japanese companies have excelled in driving flexibility, adaptability, and collaboration. Madu [136] highlights that organisational culture is vital to a company's competitive advantage and sustained performance. Leaders play a pivotal role in shaping these cultures, which influence the core values and principles of the organisation [126]. Hence, organisational culture can be understood as the shared worldview and value system that shape behaviours within a company [198]. It reflects the practices and principles embraced by individuals within the organisation [171] and determines how the organisation responds to various circumstances [179]. Stock et al. [207] describe it as the guiding philosophy for business operations, while Xue et al. [248] categorise it as a distinct form of organisational structure. Supporting this, Iroanya [106] argues that organisational culture is a decision-making framework that helps members address internal and external challenges. In addition, Sorensen [204] and Parker et al. [165] emphasise the shared beliefs of group members as a foundation for organisational behaviour, which evolves. According to Kalyar and Rafi [111], organisational culture shapes employees' perceptions of the company regarding its values, assumptions, and beliefs.

A strong organisational culture enhances competitive advantage by aligning employees with shared values and goals, streamlining business processes and driving effective implementation of strategies [250]. Such cultures attract and retain talent, instil a sense of purpose in employees, and empower decision-making across all levels. Clear cultural definitions ensure employees understand and align with long-term organisational objectives. This alignment saves time and resources, making the organisation more efficient and competitive. When integrated deeply, organisational culture transcends marketing rhetoric to become a foundational aspect of operations, holding everyone accountable to high standards and driving shared visions and expectations [191]. A strong culture also enhances resilience during crises [206]. Its critical elements (discipline, decision-making, alignment, and motivation) are essential for organisational success [232, 241]. Qi and Chau [174] identify organisational culture as a key performance indicator, linking it to developing knowledge resources vital for sustainable growth and continuity. Furthermore, organisational culture positively influences employee performance by encouraging alignment with the organisation's norms and values despite diverse individual backgrounds [238]. This cultural integration allows employees to perform efficiently and effectively [201]. Finally, the management of culture within organisations can yield diverse outcomes. While driving creativity and learning in some cases, it may lead to conflict if mismanaged [193]. Thus, understanding and cultivating the right organisational culture is crucial for enhancing performance and achieving long-term success.

5.3.3 Theories Underpinning Organisational Culture in Circular Economy Study

To explore and address the organisational culture gap in CEBM, insights are derived from social identity theory and change management (CM) theory.

5.3.4 Social Identity Theory

This study employs social identity theory to highlight the significance of organisational culture in integrating CE practices within the business models of construction firms. Social identity theory, introduced by Tajfel [213], explores how individuals behave and interact within organisations or groups. Rooted in psychology, the theory suggests that a person's self-esteem or sense of worth is closely tied to the groups or organisations they associate with. This study assumes that employees' commitment to implementing CE principles in their professional activities is influenced by their affiliation with specific construction organisations. Tajfel [213] outlines three key stages of social identity: social categorisation, social identification, and social comparison. During the social identification stage, an individual's identity is shaped by their membership in a group or organisation. This identity formation is driven by fundamental human needs such as belonging, employment, or self-esteem. At this stage, individuals may internalise the organisation's traits or culture, often prioritising them over their cultural backgrounds. Supporting this perspective, Hogg and Terry [101] argue that social identity theory allows individuals to reflect a distinct organisational culture that aligns with their group or workplace. For this study, the relevance of this theory lies in how construction firms' organisational cultures influence employees' acceptance and application of CE practices.

According to Turner [226], the social categorisation stage helps individuals classify organisations based on specific characteristics, aiding in their understanding of identity within the organisational context. This aspect of the theory aligns with Turner [228] social categorisation theory, which emphasises the importance of assigning dimensions to organisations for better comprehension of group identity. Ankrah and Langford [6] and Atuahene and Baiden [11] have built on Tajfel's [213] work to categorise organisational cultures in construction firms, emphasising the role of group identification in defining organisational identity. The final stage, social comparison, involves individuals evaluating themselves relative to other groups once they have aligned with a particular organisation [142, 213]. In this study, social comparison is utilised to analyse how different organisational cultures in construction firms influence the integration of CE principles into their business models. By leveraging the framework of social identity theory, this research aims to understand and compare the organisational approaches to CE adoption.

5.3.5 *Change Management (CM) Theory*

Transitioning to CE requires continuous efforts such as shifting paradigms, analysing actions and processes, and reevaluating structures and practices. It involves fundamentally rethinking mindsets alongside structural and procedural revisions [61, 116, 188]. Changes must be implemented across multiple organisational levels. At the firm-wide level, these include alterations to purpose, structure, and shared values,at the functional level, adjustments to procedures and systems; and at the individual level, changes in personal values, beliefs, and motivations [257]. Beyond technological or managerial system upgrades, confronting entrenched "business as usual" cultures is crucial for achieving lasting reforms and successfully adopting CE [134]. A central indicator of a paradigm shift is the evolution of organisational culture toward innovation and CE. This transition necessitates changes in cognitive frameworks, shared values, and emotional connections to organisational identity [17, 257]. For this cultural transformation to succeed and support circularity and sustainability, CM theory can provide essential guidance. CM is beneficial during the three phases of transition, old state, transition state, and new state, by addressing "soft" elements like organisational culture.

CM theory primarily focuses on building organisational competencies and delivering results by structuring change initiatives while considering factors like leadership, strategy, learning, and alignment [173]. It emphasises strategies that help individuals adapt to and accept change [119]. Before initiating a change project, organisations should assess three key variables: resources, processes, and values [46]. Resources may be tangible, such as machinery and assets, or intangible, such as knowledge and reputation. Processes refer to established task completion methods, while values guide decision-making patterns.

Kotter's model frames change as an ongoing adaptation process to shifting conditions and environments [51]. It also highlights the need for continuous employee training and education, focusing on skills like innovation, technology, material selection, and professional behaviours,for construction organisations aiming to adopt CE, creating a clear plan that integrates circular principles into their business models is essential. However, Kotter [120] notes that 70% of organisational change efforts fail because companies often lack a comprehensive strategy to achieve their objectives. Improving organisational change capacity is key to overcoming these challenges. Kotter's [120] eight-step methodology offers a structured approach to organisational change. The first step, creating a sense of urgency, is critical. Often, management fails to instil a deep, organisation-wide commitment to advancing and succeeding, which hinders progress. For construction companies seeking to integrate CE, driving a sense of urgency among employees is crucial to initiating cultural change. The second step involves forming a guiding coalition. This entails assembling a change team responsible for setting the vision, communicating effectively across the organisation, addressing obstacles, achieving short-term goals, managing ongoing projects, and establishing strategies for the newly adopted changes. Construction companies must create dedicated teams to drive the transition from traditional linear models to CEBM.

The third step involves establishing a well-defined transformation vision. This vision should explicitly outline how the future will differ from the past. Kotter [120, p. 74] identifies six crucial characteristics of a compelling vision: it must be "imaginable, desirable, feasible, focused, flexible, and communicable." Notably, a vision should be authentic and transparent to avoid employee confusion or distrust. A clear corporate vision serves as a guiding framework, steering the organisation in alignment with senior management's intentions. When urgency is harnessed to drive change, this vision becomes a catalyst for motivating individuals to act. In construction organisations transitioning to a CEBM, the vision must articulate clear objectives for the shift while introducing policies to support and sustain the transformation. The fourth step of Kotter's [120] framework emphasises the importance of effectively presenting the vision to gain organisational support. Developing robust communication strategies and plans to convey the vision to employees is critical during this stage. Messages need to be sincere and resonate emotionally, as transformation requires more than intellectual acceptance,it demands emotional commitment. Cultural shifts often evoke negative emotions such as fear, confusion, and mistrust. It is essential to mitigate these reactions and take empathetic and thoughtful actions. Demonstrating leadership through actions rather than words is particularly impactful. While encouraging dialogue is important, leading by example remains the most persuasive approach. Leaders, sponsors, and change agents within construction organisations must embody the principles of CEBM to ensure a smooth transition from a linear model to a circular one.

The fifth step focuses on enabling broad-based action by removing barriers and unlocking the organisation's full potential. Two primary obstacles can impede progress: structural limitations and operational inefficiencies. Structural constraints, such as misaligned processes or outdated workflows, hinder the organisation's ability to achieve its goals. For example, a company that claims improved productivity but retains inefficient practices faces a significant obstacle to true transformation. To address these challenges, organisations can implement performance-based incentives and conduct evaluations to align efforts with desired outcomes. In addition, equipping employees with relevant data through robust management information systems, such as competitive insights and market analysis, can empower them to perform more effectively. These measures collectively pave the way for a successful shift to CEBM.

The sixth step highlights the importance of generating short-term successes to maintain momentum and demonstrate the benefits of change. These early wins can boost employee morale and prove that the transformation is on track. However, clarifying that these initial achievements are stepping stones rather than the ultimate goal is essential. Short-term victories should not overshadow the broader objectives of cultural and systemic change. Establishing benchmarks for short- and long-term goals is vital for construction organisations. This ensures that the integration of CE principles into their business models remains on course, driving a comprehensive and sustainable transition from a linear approach to CEBM.

Kotter's [120] change management framework, step 7, "Do not let up," emphasises maintaining momentum. This phase encourages the team to leverage the benefits of early successes while striving for more significant, sustained change. Although initial victories can energise employees, resistance may still arise from others within the organisation. Therefore, managers must work to sustain and rebuild momentum continually. This involves initiating additional improvement projects, engaging new contributors to support the change effort, ensuring senior leaders remain actively involved to provide clear direction, and encouraging employees across various roles to take ownership of projects. It also includes reducing the dependency between different organisational units, maintaining focus on urgency, and consistently reinforcing evidence that integrating CE principles into business models is yielding positive results.

Step 8 approach addresses integrating the new culture within the organisation. This final stage focuses on establishing strategies that ensure lasting change. It requires ongoing transparency about the outcomes of the change process. Sharing success stories from the organisation's change initiatives and highlighting the differences between the previous and current cultures can drive greater employee openness to change. When the organisation recognises the tangible benefits of the transformation, it becomes better equipped to align with evolving management objectives and sustain the improvements achieved through the change process.

5.3.6 Dimensions of Organisational Culture

Cameron and Quinn [33] describe organisational culture as encompassing various dimensions, which act as indicators for understanding an organisation's core values. Similarly, Ankrah and Langford [6], Olanipekun et al. [160], and Nwachukwu and Chládková [157] emphasise the distinctive nature of organisational culture, highlighting its foundation in strong beliefs and clear direction. Gaining insight into an organisation's culture allows for identifying its underlying values [16, 36]. Serpell and Rodríguez [200] further argue that recognising these cultural dimensions is key to enhancing organisational performance, suggesting that understanding organisational culture can be highly advantageous for a company. O'Reilly et al. [162] outline several dimensions of organisational culture, including aggressiveness, stability, teamwork, innovation, and attention to detail. The innovation dimension, for example, emphasises an organisation's ability to adapt to emerging trends, encouraging employees to take risks and make autonomous decisions. On the other hand, a stability-oriented culture prioritises creating a secure and consistent work environment for employees.

Dimensions such as attention to detail, outcome orientation, and respect for individuals are characterised by precision, goal enforcement, and active listening. The aggressiveness dimension reflects an organisation's external posture, while teamwork focuses on internal collaboration. Cameron and Quinn [33] classify organisational culture into four main types: clan, adhocracy, hierarchy, and market cultures. Adhocracy promotes a creative work environment where employees are encouraged to take risks, while clan cultures stress teamwork and employee collaboration. Hierarchical cultures focus on structure and rule enforcement, whereas market cultures aim to meet customer demands.

Olanipekun et al. [160] utilised Sarros et al.'s [194] framework to evaluate the organisational culture within quantity surveying firms. Sarros et al. [194] identify three core components of organisational culture: the business, the people, and the external environment. The business dimension addresses the challenge of maintaining competitiveness, ensuring that employees are agile and responsive to changes in their surroundings. The workforce culture emphasises performance-based recognition rather than factors like rank or favouritism. Lastly, the external environment dimension relates to corporate social responsibility, highlighting an organisation's commitment to societal and environmental considerations.

Furthermore, Hofstede [100] identified several key cultural dimensions that shape organisational culture, with some emerging in recent years. His model includes process-oriented versus result-oriented, job-oriented versus employee-oriented, professional versus parochial, open systems versus closed systems, tight versus loose, and pragmatic versus normative characteristics. While these categories have distinct meanings, they are linked by their naming conventions. For example, the cultural distinction between process-oriented and result-oriented organisations highlights the importance of both procedures and outcomes within these organisations. In addition, Denison [54] proposed another framework for understanding organisational culture, which he derived from existing literature. The study divided organisational culture into four dimensions: adaptability, consistency, mission, and involvement. The adaptability dimension emphasises an organisation's ability to respond to shifts and trends in its external environment. Consistency ensures that all departments align with the organisation's values and objectives. The mission dimension revolves around defining the organisation's goals and the strategies employed to achieve them. Lastly, the involvement dimension promotes collaboration and teamwork within the organisation.

These cultural dimensions, as outlined by Denison and Mishra [55], have been widely discussed in frameworks such as the competing values framework (CVF) by Cameron and Quinn [33], O'Reilly et al. [162], and Sarros et al. [194]. Ankrah and Langford [6] and Gineviius and Vaitknaite [83] suggest that a study's organisational culture dimensions can be flexible, depending on the research focus. They also note that selecting widely-used dimensions from related fields is advisable. A literature review indicates that the CVF, particularly in its application to the construction industry, is the most frequently used model. Therefore, this study adopts Quinn's competing values framework to model the CEBM within the construction sector, with further details on the CVF provided below.

5.3.7 Quinn Competing Value Framework

For the organisational culture aspect of this study, the Competing Values Framework (CVF) developed by Cameron and Quinn [33], as well as the earlier work by Quinn and Rohrbaugh [175], provided the theoretical foundation. A review of existing literature indicates that this framework has been widely applied in various organisational studies within the construction industry. The CVF comprises two main dimensions and four primary clusters [33]. The first dimension differentiates between an organisation's internal focus and external orientation, often categorised into two distinct groups. The second dimension defines the intended organisational structure, subdivided into four elements: discretion, flexibility, control, and stability. These two dimensions form a four-part quadrant that highlights the key cultural characteristics, with each quadrant representing one of the following cultural types: clan, adhocracy, market, and hierarchical.

5.3.8 Clan Culture

According to Cameron and Quinn [33] and Willar et al. [242], clan culture is often likened to a family-like structure within organisations, where employees are viewed as family members. This culture values flexibility, openness, and ongoing communication. Rameezdeen and Gunarathna [177] suggested that clan culture drives employees' strong sense of dedication, encouraging them to prioritise the organisation's success over personal interests. In addition, clan culture promotes collaboration, knowledge sharing, mentorship, and teamwork. Patyal et al. [166] pointed out that organisations with a clan culture are often regarded as inclusive environments with a strong emphasis on tradition and loyalty. The study found that stakeholders in construction organisations adopt clan culture to enhance collaboration, mentorship, and partnerships within and across organisations. The CVF by Quinn and Rohrbaugh [175] emphasised that clan culture is internally focused, prioritising the engagement and commitment of stakeholders. Atuahene and Baiden [11] connected clan culture to Elton Mayo's ideas, highlighting the importance of informal relationships and collaboration in driving productivity. Albayrak and Albayrak [4] also emphasised that effective communication is key to adopting clan culture, as a lack of communication can lead to uncertainty and disrupt the informal relationships that form its foundation.

5.3.9 Adhocracy Culture

Due to the cultural influence, employees work in an adaptable environment that promotes creativity within a dynamic setting [33, 175]. The adhocracy culture focuses on the external aspect of the organisation, with innovation being the key driver of success. Organisations that adopt this culture prioritise entrepreneurial growth, employee involvement, and equitable compensation [94, 164, 177]. Leaders embodying this culture are willing to take risks, and their teams are equally committed to similar actions [11, 94]. According to Quinn and Rohrbaugh [175], adhocracy culture is especially significant for temporary organisations or businesses. Iroanya [106] further corroborated this, suggesting that it is suitable for companies that aim to complete tasks temporarily before ceasing operations. Considering the transient nature of the construction industry, this culture is seen as a good fit [130]. In addition, the adhocracy culture ensures total quality management (TQM) in the construction sector [164].

5.3.10 Market Culture

Cameron and Quinn [33] identified the market culture as an external component of organisational culture due to its focus on an organisation's external environment. Leaders who embrace this culture are often seen as strict, goal-focused, and dedicated to maximising productivity. Ankrah and Langford [6] further argued that executives who adopt a market culture are driven by production, profitability, and client satisfaction. On the other hand, employees are given clear directives regarding their duties and are compensated based on their performance [94, 256].

In relation to CEBM, this study posits that construction organisations that adopt a market culture are driven by the pursuit of client satisfaction while also considering the interests and perspectives of other stakeholders. According to Harinarain et al. [94], the competitive nature of this culture makes it a dominant organisational feature in the construction industry. Similarly, Albayrak and Albayrak [4] found that construction firms embracing this culture perform better, enabling them to focus on client satisfaction, transparent bidding, and achieving set goals. Patyal et al. [166] suggested that a construction firm can secure a significant market, drive healthy competition, and maintain a results-oriented approach by adopting a market culture. This highlights the market culture's strong emphasis on meeting the needs of various stakeholders.

5.3.11 Hierarchy Culture

This element of organisational culture focuses on maintaining a structured hierarchy and strict oversight of operations [33]. A hierarchical culture ensures that employees are categorised based on their competencies, qualifications, or tasks. This classification system establishes a chain of command that defines what, when, and how tasks are completed [11]. The characteristics of a hierarchical culture promote consistency, control, regulation of activities, and a firm adherence to policies. The findings of this study further suggest that organisations within the construction industry adopt a similar cultural approach, emphasising control, authority, and enforcement within the context of a CEBM.

5.3.12 Requirements for Circular Economy-Oriented Culture Change in Organisations

Transitioning to a Circular Economy (CE) involves several key elements that must be adjusted. These include values, mindsets, behaviours, competencies, and capabilities shifts.

5.3.13 Values

Jones [109] argued that values serve as standards or guiding principles individuals use to determine desirable or undesirable actions, situations, and outcomes. These values dictate what an organisation should or should not do and reflect its financial priorities and business model, as they establish the rules that employees must follow for the organisation to thrive [46]. Values are fundamental in driving a culture centred on CE, as they communicate the aspects the business prioritises [21]. While individuals are generally aware of their values, others may not always notice them. This study defines CE-oriented values as both a characteristic and an enabler that shapes the attitudes and behaviours necessary for transitioning to a CE.

For example, CE-oriented values play a crucial role in sustaining a culture and are integral to the future of a circular organisation. At the core of an organisation's culture lie the values that define its identity, structure, and strategy, even if these values are not immediately apparent [7, 17, 195]. Adapting and aligning these values with CE principles and strategies is essential for driving organisational change toward a CE.

5.3.14 Mindsets

Dweck [64: 625] stated that mindsets are "people's lay beliefs about the nature of human attributes […] some people hold a fixed mindset and believe that human attributes are simply fixed traits while other people hold a growth mindset and believe that their qualities can be developed through nurture and their persistent efforts". Motivation and attitude are often referred to as mindsets. According to Dweck's framework of mindsets at the organisational level, a growth mindset drives curiosity and a readiness to take risks. In contrast, a fixed mindset focuses on adhering to familiar, established methods. A mindset oriented toward the CE represents the beliefs and attitudes that align with CE principles and circular values, influencing how an organisation perceives and reacts to various situations. In this context, mindsets are considered assumptions or presumptions that guide behaviour and decision-making.

5.3.15 Behaviours

The term 'pro-circular behaviour' was created by Muranko et al. [148: 133] and is defined as "an action which is brought about due to prioritising resource-efficiency". Liu and Bai [131] suggest that internal factors and external contexts influence a firm's environmental behaviour, such as societal norms, market dynamics, and local communities. According to Joshi and Seay [110], this behaviour must be economically beneficial, environmentally sustainable, and socially responsible. Circular behaviours describe how individuals and organisations act to support the transition to CE. For this transition to succeed, leaders must inspire their employees by providing a clear vision of what the company should look like in a circular model. Employees can contribute to the CE shift by developing new competencies, as Maitre-Ekern and Dalhammar [137] argue that behaviour can be modified and cultivated to adopt a CE perspective.

Singh et al. [203] assert that individuals with behavioural intent are influenced by their attitudes, subjective norms, and perceived behavioural control. The attitudes of leaders and managers play a critical role in how CE is implemented, as these attitudes shape their cognitive and emotional states [48, 122]. Subjective norms, or the expectation that an organisation will engage in specific behaviours, are shaped by stakeholder demands [203]. Additionally, perceived behavioural control, tied to the organisation's capacity and managerial beliefs, impacts their readiness to implement CE strategies [48].

5.3.16 Capabilities

Teece [217] highlighted that capabilities refer to the ability to complete a task, encompassing operational and strategic actions. In earlier works [215, 216] introduced the idea of dynamic capabilities, which he defined as the ability to reorganise, redirect, and transform an organisation by combining existing core competencies with external resources, strategies, and complementary assets to meet the challenges of a fast-paced, competitive environment. Winter [243] described operational capability as a range of routines that, when combined with input flows, allow management to make decisions that lead to substantial outputs of a particular kind. Similarly, Teece [217] introduced the concept of "ordinary capabilities," which involve executing tasks related to administration, operations, and governance necessary for completing objectives. Teece [218: 4] further defined industrial capabilities as "able to continuously sense and seize opportunities, and to periodically transform aspects of the organisation and culture to be able to proactively reposition to address yet newer threats and opportunities as they arise".

To effectively navigate the challenges of the CE transformation, organisations need CE-specific capabilities to reconfigure, transform, redirect, shape, and integrate current core competencies with external resources, circular strategies, and complementary assets. These capabilities represent the competencies required for an organisation's transformation efforts in CE. Moreover, they reflect the new skills organisations must acquire or develop to build CEBM and systems. The knowledge created, shared, and learned through this process leads to developing new competencies, which, in turn, offer a competitive edge.

5.3.17 Competences

According to Mirabile [144: 75], competence is a "knowledge, skill, ability, or characteristic associated with high performance on a job, such as problem solving, analytical thinking, or leadership. Some definitions of competence include motives, beliefs, and values", while for Hamel and Prahalad [91: 279], "competencies are the collective learning in the organisations, especially how to coordinate diverse production skills, integrate multiple streams of technologies [...] organise the work and deliver value". Competencies are advanced capabilities that provide a competitive edge [258] and are key differentiators between organisations. Due to their strategic importance and the difficulty of replicating them, they must be continuously updated [217]. Competencies enable the consistent and efficient completion of challenging tasks and can be dynamic, allowing organisations to undertake previously unimaginable tasks [75]. With CE principles, CE-oriented competencies encompass knowledge, skills, and attitudes that enable task performance and problem-solving. These competencies reflect how healthy stakeholders (customers and competitors) perceive an organisation as circular and whether circularity has led to financial advantages.

5.3.18 Implications of Organisational Culture in Circular Economy Adoption in Construction Organisations Business Models

Organisations often hesitate to adopt policies that might impact their performance. However, considering the severe environmental degradation, society needs to shift its perspective on the environment. Business, education, and society leaders must drive the adoption of CE and sustainable practices. The common aim of these initiatives is to halt further environmental damage and conserve natural resources by effectively utilising renewable energy, managing waste, and adopting integrated CE models [66, 85, 188, 192]. A company's culture often plays a significant role in adopting new business models, such as CE. A fundamental restructuring is necessary for a company to transition to a CE, requiring complete alignment and engagement from everyone in the organisation. The success of such a transition is closely linked to organisational culture, as many studies indicate a strong correlation between culture and effectiveness [53, 121].

Organisational culture can significantly influence corporate performance and productivity [53, 96, 196]. Strong leadership, high employee performance, and an ethical worldview are critical components of a thriving culture. Corporate leaders rely on an appropriate organisational culture to execute strategies and improve effectiveness. Research [74, 161] outlines how leaders can use organisational culture to enhance productivity and performance. Culture serves as a tool for business management to shape the work environment, influencing employee attitudes, boosting operational efficiency, and enhancing financial outcomes. Through culture, organisations introduce innovative products, processes, services, and approaches that sustain internal success. Organisational culture is critical in aligning corporate strategies, especially since adopting CE is a strategic priority, making evaluating a company's cultural foundation necessary.

In a CE environment, greater employee involvement, effective leadership, and commitment to the organisation's vision and goals are crucial [44, 45]. When integrated with the CE program, these factors lead to improved performance. Successful organisations motivate their teams, drive cooperation, and continually optimise workforce potential. A cohesive and integrated organisation is more likely to succeed. It is also important for the company to ensure it can drive improvement and understand client needs while continuously gathering feedback. Viewing the organisation as a system, identifying areas for process enhancement, and improving interactions with clients and employees can lead to better performance [153, 240]. Once the organisational culture is strong enough, stakeholders will embrace innovative business models. Therefore, culture should be prioritised when implementing the CE initiative to boost corporate efficiency.

For a successful transition to CE, soft business management elements are essential. Human-centred factors, such as culture, play a crucial role in successfully adopting CE business models at the corporate level [43]. Several studies have highlighted the importance of understanding, modifying, and aligning organisational culture to facilitate CE adoption [18, 31, 107, 114]. For CE to become a transformative paradigm which fundamentally alters industrial production and consumption, norms, values, visions, tools, instruments, and indicators must be examined and adjusted [118].

5.4 Summary

This chapter examines the gaps identified in the existing literature on CEBM, focusing on the challenges of incorporating CE into the business models of construction organisations. The study highlighted the importance of material information management systems and organisational culture for successful CE integration within the construction sector. Considering the unique characteristics of the construction industry, the design and implementation of material information systems and the alignment of organisational culture are essential when discussing CE adoption. The findings of this chapter thus provide the foundation for developing the proposed framework presented in this book.

References

1. Abdul-Rahman H, Berawi MA, Berawi AR, Mohamed O, Othman M, Yahya IA (2006) Delay mitigation in the Malaysian construction industry. J Constr Eng Manag 132(2):125–133
2. Aguinis H, Glavas A (2013) Embedded versus peripheral corporate social responsibility: psychological foundations. Ind Organ Psychol 6(4):314–332
3. Ahmed M, Shafiq S (2014) The impact of organizational culture on organizational performance: a case study of telecom sector. Glob J Manage Bus Res 14(3):21–30
4. Albayrak G, Albayrak U (2014) Organizational culture approach and effects on Turkish construction sector. APCBEE Proc 9:252–257
5. Amoako-Gyampah K, Salam AF (2004) An extension of the technology acceptance model in an ERP implementation environment. Inf Manage 41(6):731–745
6. Ankrah NA, Langford DA (2005) Architects and contractors: a comparative study of organizational cultures. Constr Manag Econ 23(6):595–607
7. Ashforth BE, Mael F (1989) Social identity theory and the organization. Acad Manag Rev 14(1):20–39
8. Asif M, Searcy C, Zutshi A, Ahmad N (2011) An integrated management systems approach to corporate sustainability. Eur Bus Rev 23(4):353–367
9. Asif M, Searcy C, Zutshi A, Fisscher OA (2013) An integrated management systems approach to corporate social responsibility. J Clean Prod 56:7–17
10. Atran S (2002) In gods we trust: the evolutionary landscape of religion. Oxford University Press
11. Atuahene BT, Baiden BK (2018) Organizational culture of Ghanaian construction firms. Int J Constr Manag 18(2):177–188
12. Ayeh JK (2015) Travellers' acceptance of consumer-generated media: an integrated model of technology acceptance and source credibility theories. Comput Hum Behav 48:173–180
13. Ajzen I, Fishbein M (1980) Understanding attitudes and predicting social behaviour. Prentice-Hall, Englewood Cliffs, NJ
14. Antikainen M, Valkokari K (2016, March) Framework for sustainable circular business model innovation. In: ISPIM innovation symposium. The International Society for Professional Innovation Management (ISPIM), p 1
15. Bagozzi RP (2007) The legacy of the technology acceptance model and a proposal for a paradigm shift. J Assoc Inf Syst 8(4):3
16. Baird K, Jia Hu K, Reeve R (2011) The relationships between organizational culture, total quality management practices and operational performance. Int J Oper Prod Manage 31(7), 789–814
17. Bansal P (2003) From issues to actions: the importance of individual concerns and organizational values in responding to natural environmental issues. Organ Sci 14(5):510–527
18. Bashir M, Verma R (2019) Internal factors and consequences of business model innovation. Manag Decis 57(1):262–290
19. Bashkow I (2004) A Neo-Boasian conception of cultural boundaries. Am Anthropol 106(3):443–458

20. Bell LC, Stukhart G (1986) Attributes of materials management systems. J Constr Eng Manag 112(1):14–21
21. Bernon M, Tjahjono B, Ripanti EF (2018) Aligning retail reverse logistics practice with circular economy values: an exploratory framework. Prod Planning Control 29(6):483–497
22. Bidney D (1942) On the philosophy of culture in the social sciences. J Philos 39(17):449–457
23. Bidney D (1944) On the concept of culture and some cultural fallacies. Am Anthropologist 46(1):30–34
24. Blackmore SJ (2000) The meme machine. Oxford University Press
25. Bloch M (2012) Anthropology and the cognitive challenge, new departures in anthropology. Cambridge University Press
26. Boas F (1940) Race, language, and culture. University of Chicago Press
27. Bourdieu P (1977) Outline of a theory of practice. Cambridge University Press
28. Bourdieu P (1990) The logic of practice. Stanford University Press
29. Boyer P (2001) Religion explained: the evolutionary origins of religious thought. Basic Books
30. Brown A (1998) Organisational culture, 2nd edn. Pitman Publishing
31. Bustinza OF, Gomes E, Vendrell-Herrero F, Tarba SY (2018) An organizational change framework for digital servitization: evidence from the Veneto region. Strateg Chang 27(2):111–119
32. Caldas CH, Torrent DG, Haas CT (2004) Integration of automated data collection technologies for real-time field materials management. In: Proceedings of the 21st international symposium on automation and robotics in construction. https://www.researchgate.net/profile/CarlHaas/publication/228575275_Integration_of_automated_data_collection_technologies_for_realtime_field_materials_management/links/571d068408ae7f552a48f239/Integration-ofautomated-data-collection-echnologies-for-real-time-field-materials-management.pdf
33. Cameron KS, Quinn RE (2011) Diagnosing and changing organizational culture: based on the competing values framework. Wiley
34. Campbell D, Stonehouse G, Houston B (1999) Business strategy: an introduction. Butterworth Heinemann
35. Canter MR (1993) Resource management for construction an integrated approach. Macmillan
36. Cao Z, Huo B, Li Y, Zhao X (2015) The impact of organizational culture on supply chain integration: a contingency and configuration approach. Supply Chain Manage Int J 20(1):24–41
37. Chandler AD (1990) Strategy and structure: chapters in the history of the industrial enterpise. MIT Press
38. Chapman S, Gatewood A, Arnold T, Clive L (2016) Introduction to materials management, 8th edn. Pearson
39. Chen KM, Chen JC, Cox RA (2012) Real time facility performance monitoring system using RFID technology. Assem Autom 32(2):185–196
40. Chen W, Song X (2019) Enterprise management innovation in the internet of things from the perspective of contingency. J Intell Fuzzy Syst 37(5):5829–5836
41. Cheng JC, Ma LY (2013) A BIM-based system for demolition and renovation waste estimation and planning. Waste Manage 33(6):1539–1551
42. Chenhall RH (2003) Management control systems design within its organizational context: findings from contingency-based research and directions for the future. Acc Organ Soc 28(2–3):127–168
43. Chiappetta Jabbour CJ, Sarkis J, de Sousa L, Jabbour AB, Scott Renwick DW, Singh SK, Grebinevych O, Kruglianskas I, Filho MG (2019) Who is in charge? A review and a research agenda on the 'human side' of the circular economy. J Clean Prod 222:793–801
44. Choi I (2020) Moving beyond mandates: organizational learning culture, empowerment, and performance. Int J Public Adm 43(8):724–735
45. Choi Y, Yu Y (2014) The influence of perceived corporate sustainability practices on employees and organizational performance. Sustainability 6(1):348–364
46. Christensen CM, Overdorf M (2000) Meeting the challenge of disruptive change. Harv Bus Rev 78(2):66–77

47. Clifford J, Marcus GE (eds) (1986) Writing culture: the poetics and politics of ethnography. University of California Press
48. Corral CM (2003) Sustainable production and consumption systems—cooperation for change: assessing and simulating the willingness of the firm to adopt/develop cleaner technologies. The case of the In-Bond industry in northern Mexico. J Cleaner Prod 11(4):411–426
49. Davis FD, Bagozzi RP, Warshaw PR (1989) User acceptance of computer technology: a comparison of two theoretical models. Manage Sci 35(8):982–1003
50. Dawkins R (1976) The selfish gene. Oxford University Press
51. Dawson PM (1994) Organizational change: a processual approach. Paul Chapman Publishing
52. Deal TE, Kennedy AA (1982) Corporate cultures, The rites and rituals of corporate life. Addison-Wesley
53. Denison DR (1990) Corporate culture and organizational effectiveness. Wiley
54. Denison DR (1996) What is the difference between organizational culture and organizational climate? A native's point of view on a decade of paradigm wars. Acad Manag Rev 21(3):619–654
55. Denison DR, Mishra AK (1995) Toward a theory of organizational culture and effectiveness. Organ Sci 6(2):204–223
56. Dennett DC (1990) Memes and the exploitation of imagination. J Aesthet Art Critic 48(2):127–135
57. Dewar R, Hage J (1978) Size, technology, complexity, and structural differentiation: toward a theoretical synthesis. Adm Sci Q 111–136
58. Dey PK (2000) Managing projects in fast track–a case of public sector organization in India. Int J Public Sect Manag 13(7):588–609
59. Dilthey W (1976) Selected writings (Rickman HP, Ed.). Cambridge University Press
60. Donaldson T, Preston LE (1995) The stakeholder theory of the corporation: concepts, evidence, and implications. Acad Manag Rev 20(1):65–91
61. Dufva M, Kettunen O, Aminoff A, Antikainen M, Sundqvist-Andberg H, Tuomisto T (2016) Approaches to gaming the future: planning a foresight game on circular economy. In: Games and learning alliance: 4th international conference, GALA 2015, Rome, Italy, December 9–11, 2015, Revised selected papers 4. Springer International Publishing, pp 560–571
62. Dunaway Virginia M, Sullivan YW, Fosso Wamba S (2019) Building dynamic capabilities with the internet of things. https://doi.org/10.24251/HICSS.2019.713
63. Durham WH (1991) Coevolution: genes, culture, and human diversity. Stanford University Press
64. Dweck CS (2012) Mindsets and human nature: promoting change in the Middle East, the schoolyard, the racial divide, and willpower. Am Psychol 67(8):614
65. Ehrlich P, Feldman M (2003) Genes and cultures: what creates our behavioral Phenome? Curr Anthropol 44(1):87–107
66. Ellen MacArthur Foundation (2014) Potential for Denmark as a circular economy a case study from: delivering the circular economy—a Toolkit for policy makers. http://mst.dk/media/151170/15-11-25-cirkulaer-oekonomi.pdf
67. Ellen MacArthur Foundation, & Ide (2017) Business model canva. http://circulardesignguide.com/post/circular-business-model-canvas
68. Emmanuel O (2017) Understanding organisational culture and organisational performance: are they two sides of the same coin? J Manage Res 9(1):12–21
69. Eriksson P, Lind H (2015) Moral hazard and construction procurement: a conceptual framework. In: Procurement for sustainable innovations in the built environment. pp 1–26
70. Eslami MH, Jafari H, Achtenhagen L, Carlbäck J, Wong A (2024) Financial performance and supply chain dynamic capabilities: the Moderating Role of Industry 4.0 technologies. Int J Prod Res 62(22):8092–8109
71. European Construction Institute (2014) Total productivity management: guideline for the construction phase. http://www.eci-online.org/wp-content/uploads/2015/11/ECIARC11-Total-Productivity-Management-Guidelines-for-the-Construction-Phase.pdf

72. Fewings P (2012) Construction project management: an integrated approach, 2nd edn. Routledge
73. Finkenzeller K (2010) RFID handbook: fundamentals and applications in contactless smart cards, radio frequency identification and near-field communication, 3rd edn. Wiley
74. Flamholtz EG, Randle Y (2012) Corporate culture, business models, competitive advantage, strategic assets and the bottom line: theoretical and measurement issues. J Hum Resour Costing Account 16(2):76–94
75. Francis D, Bessant J, Hobday M (2003) Managing radical organisational transformation. Manag Decis 41(1):18–31
76. Freeman JH (1973) Environment, technology, and the administrative intensity of manufacturing organizations. Am Sociol Rev 750–763
77. Freeman RE (1984) Strategic management: a stakeholder approach. Pitman
78. Furnham A, Gunter B (1993) Corporate assessment: auditing a Company's personality. Routledge
79. Furuta H, Frangopol D, Akiyama M (eds) (2014) Life-cycle of structural systems: design, assessment, maintenance and management. CRC Press
80. Gavric G, Sormaz G, Ilic D (2016) The impact of organizational culture on the ultimate performance of a company. Int Rev 3(4):25–30
81. Geertz C (1973) The interpretation of cultures. Basic Books
82. Ghosh S, Hughes M, Hodgkinson I, Hughes P (2022) Digital transformation of industrial businesses: a dynamic capability approach. Technovation 113:102414
83. Ginevičius R, Vaitkūnaite V (2006) Analysis of organizational culture dimensions impacting performance. J Bus Econ Manag 7(4):201–211
84. Gonzalez-Perez MA, Leonard L (2015) The global compact: corporate sustainability in the post 2015 world. In: Beyond the UN global compact: institutions and regulations. Emerald Group Publishing Limited, pp 1–19
85. Govindan K, Hasanagic M (2018) A systematic review on drivers, barriers, and practices towards circular economy: a supply chain perspective. Int J Prod Res 56(1–2):278–311
86. Grieves J (2000) Introduction: the origins of organizational development. J Manage Dev 19(5):345–447
87. Griffith A, Stephenson P (2016) Management systems for construction. Chartered Institute of building. Taylor and Francis
88. Guha S, Rosenfeld JA, Malhotra AK, Lee AT, Gregersen PK, Kane JM, Pe'er I, Darvasi A, Lencz T (2012) Implications for health and disease in the geneticsignature of the Ashkenazi Jewish population. Genome Biol 13(1):R2. https://doi.org/10.1186/gb-2012-13-1-r2
89. Guo L, Wei YS, Sharma R, Rong K (2017) Investigating e-business models' value retention for start-ups: the moderating role of venture capital investment intensity. Int J Prod Econ 186:33–45
90. GUS (2020) Information society in Poland in 2020. https://Stat.Gov.Pl/Obszary-Tematy czne/Nauka-i-Technika-SpoleczenstwoInformacyjne/Spoleczenstwo-Informacyjne/Wykorz ystanie-TechnologiiInformacyjno-Komunikacyjnych-w-Jednostkach-Administracji-Public znej-Przedsiebiorstwach-i-Gosp
91. Hamel G, Prahalad CK (1996) Competing in the new economy: managing out of bounds. Strateg Manag J 17(3):237–242
92. Hamidi B, Bulbul T, Pearce A, Thabet W (2014) Potential application of BIM in cost-benefit analysis of demolition waste management. In: Construction research congress 2014: construction in a global network, pp 279–288
93. Hampton G, Baldwin AN, Holt G (2012) Project delays and cost: stakeholder perceptions of traditional v. PPP procurement. J Financ Manage Property Constr 17(1):73–91
94. Harinarain N, Bornman CL, Botha M (2013) Organisational culture of the South African construction industry. Acta Structilia 20(1):22–43
95. Harris M (1968) The rise of a history of theories of culture. Thomas Y. Crowell Company
96. Hartnell CA, Ou AY, Kinicki AJ, Choi D, Karam EP (2019) A meta-analytic test of organizational culture's association with elements of an organization's system and its relative predictive validity on organizational outcomes. J Appl Psychol 104(6):832

97. Henrich J (2015) Culture and social behavior. Curr Opin Behav Sci 3:84–89
98. Henrich J (2016) The secret of our success. Princeton University Press.https://doi.org/10.1515/9781400873296
99. Hofstede G (1980) Culture and organizations. Int Stud Manag Organ 10(4):15–41
100. Hofstede G (2011) Dimensionalizing cultures: the Hofstede model in context. Online Readings Psychol Cult 2(1):8
101. Hogg MA, Terry DJ (eds) (2014) Social identity processes in organizational contexts. Psychology Press. https://doi.org/10.4324/9781315800530
102. Holbraad M, Pedersen MA (2017) The ontological turn. Cambridge University Press
103. Hristov K (2017) Internet plus policy: a study on how China can achieve economic growth through the internet of things. J Sci Technol Policy Manage 8(3):375–386
104. Hunt V, Puglia A, Puglia M (2010) RFID: a guide to radio frequency identification. Wiley
105. Illingworth JR (2002) Construction methods and planning. CRC Press
106. Iroanya CC (2012) The relationship between organizational culture and quality of communication in project management. https://www.proquest.com/openview/7bbf9686b26c4af4fbe313820127d35c/1?pqorigsite=gscholar&cbl=18750
107. Isensee C, Teuteberg F, Griese KM, Topi C (2020) The relationship between organizational culture, sustainability, and digitalization in SMEs: a systematic review. J Clean Prod 275:122944
108. Jao-Hong C, Kuo-Liang L (2021) Enhancing the effects of service innovation in supply chains: insights from contingency theory and boundary spanning perspectives. Rev Argent Clín Psicológica 30(1):508
109. Jones GR (2013) Organizational theory, design, and change. Pearson
110. Joshi C, Seay J (2019) Building momentum for sustainable behaviors in developing regions using locally managed decentralized circular economy principles. Chin J Chem Eng 27(7):1566–1571
111. Kalyar MN, Rafi N (2013) 'Organizational learning culture': an ingenious device for promoting firm's innovativeness. Serv Ind J 33(12):1135–1147
112. Kar S, Jha KN (2020) Examining the effect of material management issues on the schedule and cost performance of construction projects based on a structural equation model: survey of Indian experiences. J Constr Eng Manag 146(9):05020011
113. Kasim N (2011) ICT implementation for materials management in construction projects: case studies. J Constr Eng Project Manage 1(1):31–36
114. Ketprapakorn N, Kantabutra S (2019) Culture development for sustainable SMEs: toward a behavioral theory. Sustainability 11(9):2629
115. Khandwalla PN (1977) Design of organizations, arcourt Brace Jovanovich. New York
116. Kjaer LL, Pigosso DC, Niero M, Bech NM, McAloone TC (2019) Product/service-systems for a circular economy: the route to decoupling economic growth from resource consumption? J Ind Ecol 23(1):22–35
117. Kong CW, Li H, Love PED (2001) An e-commerce system for construction material procurement. Constr Innov 1(1):43–54
118. Korhonen J, Honkasalo A, Seppälä J (2018) Circular economy: the concept and its limitations. Ecol Econ 143:37–46
119. Kotter JP (1995) Leading change: why transformation efforts fail. Harvard Bus Rev 59–67
120. Kotter JP (2012) Leading change. Harvard Business Press
121. Kotter J, Heskett J (1992) Corporate culture and performance. Free Press
122. Kumar B (2012) Theory of planned behaviour approach to understand the purchasing behaviour for environmentally sustainable products. In: IIMA working papers WP2012-12-08, Indian Institute of Management Ahmedabad, Research and Publication Department. https://ideas.repec.org/p/iim/iimawp/11469.html
123. Laland KN, Kumm J, Feldman MW (1995) Gene-culture coevolutionary theory: a test case. Curr Anthropol 36(1):131–156
124. Laland KN, O'brien MJ (2011) Cultural niche construction: an introduction. Biol Theory 6:191–202

125. Laubscher M, Marinelli T (2014) Integration of circular economy in business. In: Proceedings of the conference: going Green—Care Innovation 2014, Vienna, Austria, 17–20 Nov 2014
126. Lee D (2020) Impact of organizational culture and capabilities on employee commitment to ethical behavior in the healthcare sector. Serv Bus 14(1):47–72
127. Lewandowski M (2016) Designing the business models for circular economy—towards the conceptual framework. Sustainability 8(1):43
128. Lewens T (2015) Cultural evolution. Oxford University Press
129. Lewens T (2017) Human nature, human culture: the case of cultural evolution. Interface Focus 7(5):20170018. https://doi.org/10.1098/rsfs.2017.0018
130. Liu AM, Fellows RI (2008) Organisational culture of construction joint Ventures: case studies in Hong Kong. In: International conference on multi-national construction projects: "Securing high performance through cultural awareness and dispute avoidance." Shanghai, China
131. Liu Y, Bai Y (2014) An exploration of firms' awareness and behavior of developing circular economy: an empirical research in China. Resour Conserv Recycl 87:145–152
132. Liu Z, Osmani M, Demian P, Baldwin A (2015) A BIM-aided construction waste minimisation framework. Autom Constr 59:1–23
133. Lowie RH (1917) Culture and ethnology. D. C. McMurtrie
134. Lozano R (2007) Orchestrating organisational changes for corporate sustainability: overcoming barriers to change. Greener Manag Int 57:43–64
135. Lüdeke-Freund F, Gold S, Bocken NM (2019) A review and typology of circular economy business model patterns. J Ind Ecol 23(1):36–61
136. Madu BC (2012) Organization culture as driver of competitive advantage. J Acad Bus Ethics 5:1
137. Maitre-Ekern E, Dalhammar C (2019) Towards a hierarchy of consumption behaviour in the circular economy. Maastricht J Eur Comp Law 26(3):394–420
138. Marangunić N, Granić A (2015) Technology acceptance model: a literature review from 1986 to 2013. Univ Access Inf Soc 14:81–95
139. Marsh RM, Mannari H (1981) Technology and size as determinants of the organizational structure of Japanese factories. Adm Sci Q 33–57
140. Martins EC (2000) The influence of organizational culture on creativity and innovation in a university library. Unpublished Master dissertation, University of South Africa, Pretoria
141. Martins EC, Terblanche F (2003) Building organisational culture that stimulates creativity and innovation. Eur J Innov Manag 6(1):64–74
142. McLeod S (2019) Social identity theory simply psychology. https://www.simplypsychology.org/social-identity-theory.html
143. Mesoudi A (2011) Cultural evolution: how Darwinian theory can explain human culture and synthesize the social sciences. University of Chicago Press
144. Mirabile RJ (1997) Everything you wanted to know about competency modeling. Training Dev 73–77
145. Morgan G (1986) Images of organisation. Sage
146. Mory L, Wirtz BW, Göttel V (2016) Factors of internal corporate social responsibility and the effect on organizational commitment. Int J Hum Resour Manage 27(13):1393–1425
147. Muehlhausen FB (1991) Construction site utilization: impact of material movement and storage on productivity and cost. AACE Int Trans L2–1
148. Muranko Z, Andrews D, Newton EJ, Chaer I, Proudman P (2018) The pro-circular change model (P-CCM): proposing a framework facilitating behavioural change towards a circular economy. Resour Conserv Recycl 135:132–140
149. Murray M, Akhtar J (2016) Materials management with sap erp: functionality and technical configuration, 4th edn. SAP Press
150. Nasir H (2008) A Model for automated construction materials tracking. https://uwspace.uwaterloo.ca/handle/10012/4190

151. National Institute of Standards and Technology (NIST) (2012) Development and use of the NIST intelligent and automated construction job site TestbedNIST technical note 1726. https://www.researchgate.net/profile/Robert-Lipman/publication/271205757_Development_and_Use_of_the_NIST_Intelligent_and_Automated_Construction_Job_Site_Testbed/links/563cceb108ae34e98c4ac500/Development-and-Use-of-the-NIST-Intelligent-and-Automated-Construction-Job-Site-Testbed.pdf
152. Nawawi I (2013) Culture of leadership and performance organizations. Jakarta: PT. Dawn Iterpratama Mandiri 12(1)
153. Nazarian A, Atkinson P, Foroudi P (2017) Influence of national culture and balanced organizational culture on the hotel industry's performance. Int J Hosp Manag 63:22–32
154. Nguyen LD, Ogunlana SO (2004) A study on project success factors in large construction projects in Vietnam. Eng Constr Archit Manag 11(6):404–413
155. Nilsson N, Söderberg V (2015) How to future proof a business model: capture and capitalize value in the field of urban mining. Master's thesis, Blekinge Institute of Technology, Karlskrona, Sweden
156. Nnaji C, Awolusi I (2021) Critical success factors influencing wearable sensing device implementation in AEC industry. Technol Soc 66:101636
157. Nwachukwu C, Chladkova H (2019) Firm resources, strategic analysis capability and strategic performance: organisational structure as moderator. Int J Qual Res 13(1):75–94
158. Odor HO (2018) Organisational culture and dynamics. Glob J Manage Bus Res 18(1):23–29
159. Ojiako U (2012) Project failures in ICT and construction projects: a comparative study. Lap Lambert Academic Publishing
160. Olanipekun AO, Abiola-Falemu JO, Aje IO (2014) Dimensions of organisational culture in quantity surveying firms in Nigeria. Australas J Constr Econ Build 14(4):54–70
161. O'Reilly CA III, Caldwell DF, Chatman JA, Doerr B (2014) The promise and problems of organizational culture: CEO personality, culture, and firm performance. Group Org Manag 39(6):595–625
162. O'Reilly CA III, Chatman J, Caldwell DF (1991) People and organizational culture: a profile comparison approach to assessing person-organization fit. Acad Manag J 34(3):487–516
163. Osterwalder A, Pigneur Y (2010) Business model generation: a handbook for visionaries, game changers, and challengers, vol 1. Wiley
164. Panuwatwanich K, Nguyen TT (2017) Influence of organisational culture on total quality management implementation and firm performance: evidence from the Vietnamese construction industry. Manage Prod Eng Rev 8(1):5–15
165. Parker D, Lawrie M, Hudson P (2006) A framework for understanding the development of organisational safety culture. Saf Sci 44(6):551–562
166. Patyal VS, Ambekar S, Prakash A (2020) Organizational culture and total quality management practices in Indian construction industry. Int J Product Perform Manag 69(5):895–913. https://doi.org/10.1108/IJPPM-10-2018-0368
167. Payne AC, Chelsom JV, Reavill LR (2005) Management for engineers, 2nd edn
168. Peters M, Ribeiro A, Oseyran J, Wang K (2017) Buildings as material banks and the need for innovative business models. https://www.bamb2020.eu/topics/businessmodels/
169. Perera S, Ingirige B, Ruikar K, Obonyo E (2017) Advances in construction ICT and e-business. Routledge
170. Porter ME, Kramer MR (2006) The link between competitive advantage and corporate social responsibility. Harv Bus Rev 84(12):78–92
171. Pothukuchi V, Damanpour F, Choi J, Chen CC, Ho Park S (2002) National and organizational culture differences and international joint venture performance. J Int Bus Stud 33:243–265
172. Pratono AH (2016) Strategic orientation and information technological turbulence: contingency perspective in SMEs. Bus Process Manag J 22(2):368–382
173. PROSCI (2018) Definition of change management. https://www.prosci.com/resources/articles/change-management-defi
174. Qi C, Chau PYK (2018) Will enterprise social networking systems promote knowledge management and organizational learning? An empirical study. J Organ Comput Electron Commer 28(1):31–57

175. Quinn RE, Rohrbaugh J (1981) A competing values approach to organizational effectiveness. Public Prod Rev 122–140
176. Radim P (1933) The method and theory of ethnology. South Hadley
177. Rameezdeen R, Gunarathna N (2003) Organisational culture in construction: an employee perspective. Constr Econ Build 3(1):19–30
178. Ramkumar M, Jenamani M (2014) Sustainability in supply chain through e-procurement—an assessment framework based on DANP and liberatore score. IEEE Syst J 9(4):1554–1564
179. Ravasi D, Schultz M (2006) Responding to organizational identity threats: exploring the role of organizational culture. Acad Manag J 49(3):433–458
180. Rehman SU, Mohamed R, Ayoup H (2019) The mediating role of organizational capabilities between organizational performance and its determinants. J Glob Entrep Res 9(1):1–23
181. Reid GC, Smith JA (2000) The impact of contingencies on management accounting system development. Manag Account Res 11(4):427–450
182. Ren Z, Sha L, Hassan TM (2007) RFID facilitated construction material management-A case study of water supply project. In: Proceedings of the 24th CIB W78 conference information technology in construction, pp 401–406
183. Rebelo M, Santos G, Silva R (2014) Conception of a flexible integrator and lean model for integrated management systems. Total Qual Manag Bus Excell 25(5–6):683–701
184. Richerson PJ, Boyd R (2004) Not by genes alone. University of Chicago Press
185. Richerson PJ, Boyd R (2013) Rethinking paleoanthropology. Evol Mind, Brain, Cult 263–302
186. Risjord M (2007) Ethnography and culture. In: Philosophy of anthropology and sociology. Elsevier, pp 399–428
187. Risjord MW (2012) Models of culture. In: Kincaid H (ed) The Oxford handbook of philosophy of social science. Oxford University Press, pp 387–408
188. Ritzén S, Sandström GÖ (2017) Barriers to the circular economy–integration of perspectives and domains. Procedia Cirp 64:7–12
189. Robinson H, Udeaja C (2015) Reusing knowledge and leveraging technology to reduce design and construction costs. In: Design economics for the built environment: impact of sustainability on project evaluation. pp 227–239
190. Rosaldo R (1993) Culture and truth: the remaking of social analysis. Boston
191. Sadri G, Lees B (2001) Developing corporate culture as a competitive advantage. J Manage Dev 20(10):853–859
192. Saidani M, Yannou B, Leroy Y, Cluzel F, Kendall A (2019) A taxonomy of circular economy indicators. J Clean Prod 207:542–559
193. Samaraweera A, Senaratne S (2012) Understanding project culture in construction: a literature synthesis. http://Dl.Lib.Uom.Lk/Handle/123/17004
194. Sarros JC, Gray J, Densten IL, Cooper B (2005) The organizational culture profile revisited and revised: an Australian perspective. Aust J Manag 30(1):159–182
195. Schein EH (1984) Coming to a new awareness of organizational culture. Sloan Manage Rev 25(2):3–16
196. Schein EH (2010) Organizational culture and leadership, 4th edn. Jossey-Bass
197. Schneider SC, Barsoux JL (1997) Managing across cultures. Hall
198. Schneider B, Ehrhart MG, Macey WH (2013) Organizational climate and culture. Annu Rev Psychol 64(1):361–388
199. Scott JT (2017) The sustainable business: a practitioner's guide to achieving long-term profitability and competitiveness. Routledge
200. Serpell A, Rodriguez D (2002) Studying the organisational culture of construction companies: a proposed methodology. CIB Rep 1(1):76–91
201. Shahzad F, Luqman RA, Khan AR, Shabbir L (2012) Impact of organizational culture on organizational performance: an overview. Interdisc J Contemp Res Bus
202. Silva R, Bernardo CDP, Watanabe CYV, Silva RMPD, Neto JMDS (2020) Contributions of the internet of things in education as support tool in the educational management decision-making process. Int J Innovation Learn 27(2):175–196

203. Singh MP, Chakraborty A, Roy M (2018) Developing an extended theory of planned behavior model to explore circular economy readiness in manufacturing MSMEs, India. Resour Conserv Recycl 135:313–322
204. Sørensen JB (2002) The strength of corporate culture and the reliability of firm performance. Adm Sci Q 47(1):70–91
205. Sperber D (1996) Explaining culture. Blackwell Publishing
206. Starbuck WH (2017) Organizational learning and unlearning. Learn Organ 24(1):30–38
207. Stock GN, McFadden KL, Gowen CR III (2007) Organizational culture, critical success factors, and the reduction of hospital errors. Int J Prod Econ 106(2):368–392
208. Stukhart G (2007) Materials management approach for small scale sector. Marcel Dekker Inc
209. Sullivan G, Barthorpe S, Robbins S (2011) Managing construction logistics. Wiley
210. Sun M, Howard R (2004) Understanding I.T. in construction. Spoon Press
211. Swinburne J, Udeaja C, Tait N (2010) Measuring material wastage on construction sites: a case study of local authority highway projects. Built Nat Environ Res Papers 3(1):31–41
212. Syakur N (2011) Manajemen organisasi. Cita Pustaka MediaPerintis. In: Tabachnick BG, Fidell LS (2001) Using multivariate statistics, 4th edn. Allyn & Bacon
213. Tajfel H (1972) Social categorization, English manuscript of La categorization sociale. Introduction Psychol Soc 1. https://cir.nii.ac.jp/crid/1573387450700299392.bib?lang=en
214. Tan HC, Anumba CJ, Carrillo PM, Bouchlaghem NM, Kamara JM, Udeaja CE (2010) Capture and reuse of project knowledge in construction. Wiley-Blackwell Publication
215. Teece DJ (2000) Strategies for managing knowledge assets: the role of firm structure and industrial context. Long Range Plan 33(1):35–54
216. Teece DJ (2010) Business models, business strategy and innovation. Long Range Plan 43(2–3):172–194
217. Teece DJ (2014) The foundations of enterprise performance: dynamic and ordinary capabilities in an (economic) theory of firms. Acad Manag Perspect 28(4):328–352
218. Teece DJ (2017) Towards a capability theory of (innovating) firms: implications for management and policy. Camb J Econ 41(3):693–720
219. Teece DJ, Pisano G (1994) The dynamic capabilities of firms: an introduction. Ind Corp Chang 3(1):537–556
220. Teece DJ, Pisano G, Shuen A (1997) Dynamic capabilities and strategic management. Strateg Manag J 18(7):509–533
221. Temiz I, Calis G (2017) Selection of construction equipment by using multi-criteria decision making methods. Procedia Eng 196:286–293
222. Thaler A, Ash E, Gan-Or Z, Orr-Urtreger A, Giladi N (2009) The LRRK2 G2019S mutation as the cause of Parkinson's disease in Ashkenazi Jews. J Neural Transm 116:1473–1482
223. Thunberg M (2016) Developing a framework for supply chain planning in construction, vol 1782. Linköping University Electronic Press
224. Toktaş-Palut P, Baylav E, Teoman S, Altunbey M (2014) The impact of barriers and benefits of e-procurement on its adoption decision: an empirical analysis. Int J Prod Econ 158:77–90
225. Tompkins JA, White JA, Bozer YA, Tanchoco JMA (2010) Facilities planning, 4th edn. Wiley
226. Turner JH (1985) In defense of positivism. Soc Theory 3(2):24–30
227. Turner S (1994) The social theory of practices. University of Chicago Press
228. Turner JC (2010) Social categorization and the self-concept: a social cognitive theory of group behaviour. In: Postmes T, Branscombe NR (eds) Rediscovering social identity. Psychology Press, pp 243–272
229. Tylor EB (1889) Primitive culture: researches into the development of mythology, philosophy, religion, art and custom (J Murray ed). Henry Holt and Company. https://doi.org/10.1037/12987-000
230. Udeaja C, Ekundayo D, Zhou L, Perera S (2013) Material waste in the construction industry: a review of the legislative and supply chain issues. In: Reuse of materials and byproducts in construction: waste minimization and recycling. pp 5–27
231. Umrani WA, Shah SMM, Memon PA, Samo AH (2017) Organizational culture and business performance: an empirical investigation in the Pakistani context. Int J Acad Res Econ Manage Sci 6(1):93–107

232. Urban B, Gaffurini E (2018) Social enterprises and organizational learning in South Africa. J Entrepreneurship Emerg Economies 10(1):117–133
233. Valero E, Adán A, Cerrada C (2015) Evolution of RFID applications in construction: a literature review. Sensors 15(7):15988–16008
234. Valkokari K, Valkokari P, Palomäki K, Uusitalo T, Reunanen M, Macchi M et al (2014) Road-mapping the business potential of sustainability within the European manufacturing industry. Foresight 16(4):360–384
235. Van de Ven AH, Drazin R (1985) The concept of fit in contingency theory. Res Organ Behav 7(4):333–365
236. Van de Ven AH, Ferry DL (1980) Measuring and assessing organizations. Wiley
237. Venkatesh V, Davis FD (2000) A theoretical extension of the technology acceptance model: four longitudinal field studies. Manage Sci 46(2):186–204
238. Vlaicu FL, Neagoe A, Ţîru LG, Otovescu A (2019) The organizational culture of a major social work institution in Romania: a sociological analysis. Sustainability 11(13):3587
239. Vrat P (2014) Materials management. Springer Texts Bus Econ 10:978–981
240. Warrick DD (2017) What leaders need to know about organizational culture. Bus Horiz 60(3):395–404
241. Wetzel R, Tint B (2019) Using applied improvisation for organizational learning in the Red Cross Red Crescent climate centre. In: Sensuous learning for practical judgment in professional practice: volume 2: arts-based interventions. pp 47–73
242. Willar D, Trigunarsyah B, Coffey V (2016) Organisational culture and quality management system implementation in Indonesian construction companies. Eng Constr Archit Manag 23(2):114–133
243. Winter SG (2000) The satisficing principle in capability learning. Strateg Manag J 21(10–11):981–996
244. Won J, Cheng JC, Lee G (2016) Quantification of construction waste prevented by BIM-based design validation: case studies in South Korea. Waste Manage 49:170–180
245. Wong ET, Norman G (1997) Economic evaluation of materials planning systems for construction. Constr Manag Econ 15(1):39–47
246. Wood CR, Alvarez MW (2005) Emerging construction technologies: a FIATECH catalogue. https://nap.nationalacademies.org/read/23126/chapter/12
247. Woods M (2009) A contingency theory perspective on the risk management control system within Birmingham City Council. Manag Account Res 20(1):69–81
248. Xue X, Shen Q, Ren Z (2010) Critical review of collaborative working in construction projects: Business environment and human behaviors. J Manag Eng 26(4):196–208
249. Yadav S, Luthra S, Garg D (2021) Modelling Internet of things (IoT)-driven global sustainability in multi-tier agri-food supply chain under natural epidemic outbreaks. Environ Sci Pollut Res 28(13):16633–16654
250. Yilmaz C, Ergun E (2008) Organizational culture and firm effectiveness: an examination of relative effects of culture traits and the balanced culture hypothesis in an emerging economy. J World Bus 43(3):290–306
251. Zahid M, Ghazali Z (2015) Corporate sustainability practices among Malaysian REITs and property listed companies. World J Sci Technol Sustain Dev 12(2):100–118
252. Zahid M, Rahman HU, Mirza MZ, Memon MA (2019a) Workplace sustainability in emerging economies: A comprehensive evidence from Malaysia. City Univ Res J 9(3):497–512
253. Zahid M, Rahman HU, Muneer S, Butt BZ, Isah-Chikaji A, Memon MA (2019b) Nexus between government initiatives, integrated strategies, internal factors and corporate sustainability practices in Malaysia. J Clean Prod 241:118329. https://doi.org/10.1016/j.jclepro.2019.118329
254. Zahle J (2017) Ability theories of practice and turner's criticism of Bourdieu. J Gen Philos Sci 48:553–567
255. Zahra SA, Sapienza HJ, Davidsson P (2006) Entrepreneurship and dynamic capabilities: a review, model and research agenda. J Manage Stud 43(4):917–955

256. Zheng W (2005) The impact of organizational culture, structure, and strategy on knowledge management effectiveness and organizational effectiveness. University of Minnesota. https://www.proquest.com/openview/7cce082d60e6a3d1119062b037b57e7b/1?pqorigsite=gscholar&cbl=18750&diss=y
257. Zollo M, Cennamo C, Neumann K (2013) Beyond what and why: understanding organizational evolution towards sustainable enterprise models. Organ Environ 26(3):241–259
258. Zott C (2003) Dynamic capabilities and the emergence of intraindustry differential firm performance: insights from a simulation study. Strateg Manag J 24(2):97–125

Chapter 6
Conceptualised Circular Economy Business Model Adoption Framework for Construction Organisations

Abstract CEBMs are often recognised as tools for generating value by recycling resources and materials traditionally considered waste after their functional lifespan. Hence, this chapter presents a conceptual framework for adopting CEBM in construction organisations. It identifies twelve core latent variables that define its structure: value proposition, client segment, channels, client relationship, key resources, key activities, key partnerships, cost structure, revenue streams, takeback system, organisational culture, and material information management system. These variables were determined through findings from an extensive literature review. In addition, the chapter emphasises the framework's role in illustrating the rationale behind selecting these variables to construct an integrated conceptual CEBM.

6.1 Introduction

The components of Circular Economy Business Models (CEBMs) are rooted in the principles of the circular economy [21]. Multiple frameworks exist to derive CEBMs, but one of the most widely recognised in literature is the ReSOLVE framework, introduced by the Ellen MacArthur Foundation [8]. ReSOLVE encompasses six key elements (Regenerate, Share, Optimise, Loop, Virtualise, and Exchange), which outline the business activities and opportunities essential for applying circular economy principles. The "regenerate" principle emphasises replenishing biological resources extracted from ecosystems and promotes using renewable energy. "Share" focuses on maximising product utilisation by enabling simultaneous or sequential use by different users, thus facilitating reuse. The "optimise" principle aims to enhance product efficiency and reduce waste during production. "Loop" ensures the retention of materials within a closed-loop system, minimising resource loss. "Virtualis" advocates for delivering services and utilities digitally rather than through physical means. Lastly, "exchange" involves replacing outdated materials with advanced alternatives.

Furthermore, a thorough understanding of CEBMs requires examining the broader concept of business models (BMs). According to Zott et al. [41], a business model outlines the process of creating and capturing value. Teece [35] describes a BM as

© The Author(s), under exclusive license to Springer Nature Switzerland AG 2025
O. Otasowie et al., *Circular Economy Business Model for Construction Organisations*, Green Energy and Technology, https://doi.org/10.1007/978-3-031-88322-4_6

the "design or architecture of the value creation, delivery, and capture mechanisms employed." This perspective highlights the importance of understanding client needs, determining their willingness to pay, and designing effective mechanisms to deliver value, secure payments, and achieve profitability through the value chain. Osterwalder and Pigneur [26] similarly define a BM as "the rationale of how an organisation creates, delivers, and captures value." The study further developed the Business Model Canvas, which identifies the core building blocks of a BM. Therefore, Linder and Williander [22: 2] specified the CEBM as

> a business model in which the conceptual logic for value creation is based on utilising economic value retained in products after use in the production of new offerings. Thus, a circular business model implies a return flow to the producer from users, though there can be intermediaries between the two parties. The term circular business model, therefore, overlaps with the concept of closed-loop supply chains, and always involves recycling, remanufacturing, reuse or one of their sibling activities (e.g. refurbishment, renovation, and repair).

Hence, CEBMs are often recognised as tools for generating value by recycling resources and materials traditionally considered waste after their functional lifespan. To conceptualise a CEBM, selecting a BM framework from the existing literature is essential. One commonly used approach is the broad BM framework with nine dimensions developed by Osterwalder and Pigneur [26], which has been widely applied in CE research [9, 21] to design CEBM canvases. Lewandowski [21] CEBM canvas emphasises material loops, a key principle of the circular economy. This framework operates under the premise that products, their components, and materials can be retrieved from consumers through reverse logistics and then cascaded (for biological nutrients), reused, redistributed, remanufactured, refurbished, or recycled (for technical nutrients). Reverse logistics aligns with CE principles by incorporating practices such as take-back programs, incentivised returns, reuse initiatives, and the systematic collection of outdated goods. Meanwhile, the CEBM canvases developed by EMF and IDEO [9] and Lüdeke-Freund et al. [24] highlight a shift in business strategies. These models focus on transitioning from selling physical products to offering services as the primary value proposition. The studies also emphasise delivering digital products, reducing waste, and driving strong, lasting relationships with users and clients, among other strategies.

Thus, the present study asserts that the core components of the CEBM include value proposition (20 variables), client segment (7 variables), channels (13 variables), client relationship (13 variables), key resources (7 variables), key activities (13 variables), key partnerships (13 variables), cost structure (25 variables), revenue streams (20 variables), and takeback system (12 variables). These elements have been identified through a comprehensive literature review examining variables relevant to the CEBM. Most existing CEBM research commonly incorporates these aspects. However, this book also emphasises the significance of material information management systems (13 variables) and organisational culture (23 variables), which are explored in depth. The ten foundational constructs shaping the CEBM framework are elaborated in the subsequent section of this chapter, along with the two newly introduced constructs.

6.1.1 Value Proposition (VP)

A VP is a strategic concept businesses use to communicate how they aim to deliver value to clients or consumers. According to Webster [39], a VP should serve as the organisation's primary guiding principle, as it is fundamental to the process of creating value [27] and significantly influences overall performance. In addition, VP is among the most frequently discussed concepts in business [1]. In CEBMs, the VP is a critical component. De Jong et al. [6] highlight that it can be integrated with products, services, or a combination of both. Circular products enable product life extension through mechanisms like repair, maintenance, refurbishment, or resale [29, 38]. These products are created with materials that ensure they can be reused or recycled, thereby eliminating waste at the end of their lifecycle.

Another approach, the product-service system, allows companies to provide access to their products while retaining ownership, contrasting with traditional models where consumers purchase and own items. Examples of this system include Philips' pay-per-light service [20] and the Power-by-the-Hour model for jet engines [3]. Pearce [28] underscores the importance of tailoring VPs to meet the needs of specific user groups. Collaborative consumption models such as product rental, sharing, and pooling contribute to cost savings and customised client services. Examples of sharing-based VPs include flexible seating, shared parking spaces, tool-sharing platforms, co-living arrangements, and shared workspaces, which often require specialised digital platforms [8, 16, 38]. Businesses may offer incentives, such as buy-back schemes, to encourage consumer participation [40]. These schemes create customer value while enabling businesses to recover valuable materials, components, or products. Table 6.1 summarises the VP features proposed for developing a CE business model for construction organisations.

6.1.2 Client Segment (CS)

Paying customers or consumers with similar needs and traits are categorised as client segments (CS) [19]. Client segmentation can take various forms, including niche, continuum-segmented, diverse, and multi-sided approaches. Although a product or service may have multiple target markets, these are often prioritised. When entrepreneurs deliberately select and focus on specific segments, this process forms the backbone of the company's corporate strategy and marketing approach [30]. Alternatively, some companies may opt to position themselves as low-cost providers. The client segments block represents the various groups of individuals or organisations a business seeks to serve. Understanding and defining these segments is vital in creating a thriving CEBM. For this study, client segment attributes are summarised in Table 6.1. These attributes include quality-conscious clients, cost-conscious clients, environmentally conscious clients, business-to-business clients, business-to-client suppliers, business-to-business suppliers, and clients-to-client suppliers.

Table 6.1 Constructs and measurement variables

Latent constructs	Measurement variables
Value proposition (VP)	Refurbishing construction components/items Remanufacturing construction components/items Recycling construction components/items Using material waste like concrete as a production input Manufacturing long-lasting construction components/item Building maintenance Construction component/item renting Construction component/item leasing Producing construction components/items only on demand Constructing with fewer resources and less waste Facilitating collaboration among stakeholders Building upgrading Designing out waste Facilitating client education Facilitating takeback of used construction items/materials Construction waste handling Upcycling construction component/items Downcycling construction components/items Construction waste processing Repairing construction component/item
Client segment (CS)	Quality conscious-clients Cost-conscious clients Green clients Business-to-business clients Business-to-clients suppliers Business-to-business suppliers Clients-to-clients suppliers
Channels (CH)	Word-of-mouth Direct personal selling Wholesalers Retailers Websites Web advertisements Emails Distributors Social media Video conferences Newsletters Directories/listing Electronic-learning

(continued)

Table 6.1 (continued)

Latent constructs	Measurement variables
Client relationship (KR)	Client identification Target clients analysis Client segmentation Client attraction Direct marketing Web advertisements Emails Social media Newsletters Electronic learning Client retention One-to-one marketing Complaint management Client development Client lifetime value Upselling/cross-selling Market basket analysis
Key resources (KR)	Circular sourcing Substitution of resources with better performing materials Direct visualisation of materials (digitalization) Energy from renewable sources Operating in efficient buildings Sustainable production locations Land restoration or reclamation
Key activities (KA)	Increasing performance Housekeeping Process control Equipment modification and technology changes Sharing and virtualization—desk sharing, audio and video conferences Produce design Less raw material or energy Prolonging product/component life Resource life extension Others Remanufacturing components Recycling components Lobbying Taking back used products, components or materials Providing used products, components or materials Providing access to a product functionality

(continued)

Table 6.1 (continued)

Latent constructs	Measurement variables
Key partnerships (KP)	Suppliers Manufacturers Wholesalers Retailers Service providers Public institutions Collectors of products, components and material waste Project managers Architects Engineers Construction managers Facilities managers Waste disposal managers
Cost structure (COS)	Cost of waste handling Cost of waste processing Cost of repair Cost of maintenance Cost of remanufacturing Cost of resource inputs Cost of transportation and logistics Cost of supply risk Cost of labour Cost of facilitating collaboration Cost of take-back management Cost of client education Cost of upgrading Cost of designing out waste Cost of refurbishing products/components Cost of recycling products/components Cost of energy recovery Cost of producing long-lasting products/components Cost of materials waste use for production input Cost of chemical leasing Cost of product/component renting Cost of product/component leasing Cost of producing on demand Cost of lean thinking Cost of products/components upcycling Cost of products/components downcycling

(continued)

Table 6.1 (continued)

Latent constructs	Measurement variables
Revenue stream (RS)	Product service systems Pay-per-service/product/component Availability based (subscription) Usage based (pay per use) Performance based (solution oriented, demand fulfilment oriented) Outsourcing Others (payment received for) Product/component lease Product/component renting Building maintenance Repair of products/components Building upgrading Remanufacturing components Transformation Waste handling and processing Client education Designing out waste Refurbishing products/components Recycling products/components Producing on demand Products/components upcycling Products/components downcycling
Takeback system (TBS)	Incentivised return Collection of used products Volumes of products/materials returned Ease of take-back Transportation system Distance of take-back Condition of returned material/product Documentation Recovery rate Client satisfaction with system Environmental impact of system Cost effectiveness of system

(continued)

Table 6.1 (continued)

Latent constructs	Measurement variables
Material information management (MIMS)	Standardisation of all materials related data Centralization of all materials related data Monitoring and managing inventory levels in real time Materials requirements planning Creation of bills of materials Management of bills of materials Suppliers' information management Suppliers' relationship management Suppliers' performance management Tracking suppliers compliance with quality and regulatory requirements Reporting and analytics Integration with other systems Material traceability

(continued)

Table 6.1 (continued)

Latent constructs	Measurement variables
Organisational culture (OC)	Stakeholders' involvement in CEBM Cooperation to achieve CE Reward and incentive system to achieve CE Efficiency among project team members to achieve CE Meeting CE targets with adherence to schedules CE entrepreneurship development Regulation of CEBM activities Assertiveness on CEBM policies Client satisfaction to CEBM outcome CE result-oriented Teamwork to achieve CE CE mentoring Leadership control to achieve CE Smooth CEBM activity operations Prediction of activity with the CEBM CEBM skills development and and acquisition Emphasis on CE goal accomplishment Acquisition of new technology that supports CE CE innovation Ensuring a large market size Motivation to achieve CE Updated technology systems that support CE Healthy competition
CEBM outcome	Profitability Variable cost reduction Sustainable innovation Economic gain Extended producer responsibility Increased revenues Increased market share Economic efficiency Environmental gain Eco-friendly labelling Raw materials reduction Increased employment Improved energy efficiency Quick response to market demand Resource optimisation Waste reduction Water effluents reduction Community development Promote industrial upgrading Connectivity of material streams Collaboration among stakeholders Improved productivity Air emissions reduction

6.1.3 Channel (CH)

Virtualisation represents a significant shift toward adopting a CEBM, particularly in the context of channels. This approach enables companies to provide value propositions digitally through virtual platforms or by selling physical goods via online marketplaces [8]. In addition, businesses can engage with customers through virtual communication methods such as online advertising, email campaigns, websites, social media platforms, or video conferencing tools [8]. Therefore, channels encompass a company's various methods to communicate with customers or deliver its value proposition.

6.1.4 Client Relationship (CR)

The client relationship is a crucial aspect of a company's framework, defining how the business engages with its clients at different points of their journey. This involves the strategies and methods to attract, maintain, and nurture connections with key client segments. In the context of the CE, which prioritises waste elimination, client relationships can play a vital role in two main ways: producing items based on demand and enabling clients to select their desired products [38]. In addition, implementing Recycling 2.0 may enhance social marketing initiatives and strengthen collaborations with local partners [15]. Table 6.1 provides a detailed overview of the attributes of the client relationship considered in this book.

6.1.5 Key Resources (KR)

The key resources required to create, market, and deliver value propositions through specific channels and build and sustain relationships while generating income align with the principles of the CE in two key ways. First, they emphasise the careful selection of inputs, and second, they support the restoration and regeneration of natural capital. Input selection involves modifying products and materials to align with CE goals. This can be achieved through circular sourcing, which relies on materials and products derived from closed material loops, as outlined in four circular flows [17, 29, 38]. Another approach is directly replacing resources with higher-performing alternatives that meet or exceed technical requirements while being less environmentally harmful and more practical [7]. Key resource attributes considered for this study are summarised in Table 6.1.

6.1.6 Key Activities (KA)

The principles of the CE can be applied in various ways across activities that contribute to the creation, delivery, and enhancement of value propositions, either directly or indirectly. Some efforts are directed toward remanufacturing, recycling, or advocacy, while others focus on improving performance, refining product design, and facilitating technology exchange. Proactive maintenance, advanced process control, technological upgrades, equipment optimisation, sharing, and virtualisation can significantly boost performance. Practical housekeeping and process control involve eliminating inefficiencies, such as leaks, spills, and overheating, while ensuring optimal conditions for critical parameters like pH, temperature, pressure, water levels, and timing [7]. This approach is supported by lean thinking and management, requiring continuous monitoring, preventive maintenance programs, heightened employee environmental awareness, and incentive systems [7, 32]. A notable recent trend is the "bring-your-own-device" (BYOD) strategy, which assumes users will bring their devices to access services. This reduces the demand for additional items. For example, Citrix rewards employees for using personal computers to connect to corporate networks from home [40].

Technological advancements and equipment upgrades also play a crucial role in enhancing manufacturing processes. These improvements boost the efficiency of material, water, and energy usage, minimise emissions, and eliminate hazardous substances from production [7]. One illustrative example is using 3D printing to produce essential items [38]. Emerging technologies such as the Internet of Things, big data, machine learning, and virtualisation further enhance efficiency. Approaches like teleworking, desk-sharing, flexible seating, office hoteling, teleconferencing, and virtual office setups contribute to resource optimisation and better performance [8, 16, 31]. A comprehensive overview of the key activities analysed in this study is provided in Table 6.1.

6.1.7 Key Partnerships (KP)

Businesses can derive numerous advantages from cooperative networks, such as support in research, product development, marketing, administrative tasks, supply chain management, financial operations, manufacturing processes, and overall management [14, 32]. Companies can more effectively secure critical resources and execute essential functions by driving collaboration. For instance, industrial waste recycling is often outsourced to specialised third-party companies, which either process waste for resale to other industries or recycle materials after their initial use [7]. Through collaborative production efforts, materials can be maintained within a "closed material loop," which relies on partnerships spanning the entire value chain [38]. Sheu [33] emphasises that collaborative relationships are vital for the success of green supply chains. In the construction industry context, the more circular practices adopted by stakeholders, the greater the industry's circularity. Table 6.1 outlines the key partnership attributes analysed in this study.

6.1.8 Cost Structure (COS)

The cost structure component highlights a business's primary expenses while implementing a CEBM. Understanding and managing these costs is essential for creating a sustainable and profitable business. Adjustments to the cost structure may require organisational changes, such as modifications to staff behaviour, energy consumption, and material use. These changes can also lead to more circular shifts in the business model, with evaluating the cost structure often serving as the initial step in this process. Factors related to the cost structure can help assess the success of optimisation strategies [20, 25]. In discussions about the benefits and consequences of CE, cost structure frequently arises, often linked to investments [4], production costs, product development expenses [5], or potential cost reductions from product service systems or reverse material flows [10, 12, 13, 34]. A detailed overview of the cost structure elements addressed in this study is presented in Table 6.1.

6.1.9 Revenue Stream (RS)

A business's revenue streams refer to how it generates income. Circular business models focus on product-service systems [37, 38]. The first type, pay-per-product or service (PSS), is input-based. Availability-based PSS includes two models: progressive purchasing, where customers make incremental payments before a more significant purchase, and subscription-based renting, where users pay a regular, lower fee to access a product or service. Another model is usage-based PSS, or pay-per-use, where customers pay a one-time fee to use a product or service. The fourth model is contract- or performance-based, similar to performance leasing. Traditional income generation methods, such as selling pure goods or services, are also widely used [36]. The value proposition of the business influences the revenue streams. They may also be linked to value recovery from retrieved products, parts, or raw materials. For example, recovered product components might be resold in their original form, remanufactured, or, if valuable enough, used to create new products [18, 29]. The value must be sufficiently high or low to make material recovery economically feasible. In addition, the energy recovered from waste disposal can impact the overall value recovery [8]. A summary of the revenue stream characteristics considered in this study is presented in Table 6.1.

6.1.10 Takeback System (TBS)

The central principle of the CE is the creation of material loops [8, 25]. This principle is based on the idea that products, components, or materials can be recycled, remanufactured, refurbished, reused, redistributed, or cascaded (for biological nutrients). To support these processes, reverse logistics and collection systems are essential [8, 11, 16, 38]. Reverse logistics can be enhanced by CE practices such as take-back management, return and reuse incentives, and strategies for collecting used goods. These activities may also require the involvement of new partners, channels, and customer relationships. A summary of the revenue streams examined for this study is presented in Table 6.1.

6.2 Model Justification

This book aims to develop a CEBM tailored for construction organisations. As discussed in earlier sections, it is evident that a BM involves creating, delivering, and capturing value and is a key source of competitive advantage. At its core, a BM is designed to identify client needs and their willingness to pay. In addition, the foundational elements of a BM were identified as key components in all the CEBMs reviewed in Chap. 3. These elements serve as a guide to help organisations organise their efforts to create value, generate profits, and achieve success. However, the BM, central to the development of the CEBMs discussed, was adapted to incorporate CE principles. This adaptation was made by integrating CE principles into the BM's core elements to reduce the wasteful consumption of natural resources. This shift allows companies to gain a competitive edge while driving a strong relationship with the environment. While the models of Lüdeke-Freund et al. [23], EMF and IDEO [9], and Antikainen and Valkokari [2] all apply CE principles to BM core elements, Lewandowski [21] CEBM stands out as the most comprehensive. The study's model emphasises a closed-loop production process, where materials or products typically discarded are recovered through a take-back system. These recovered materials are either reused in production or composited to provide environmental nutrients. Hence, the conceptualisation of the CEBM by Lewandwoski [21] is fundamental to developing the framework in this book.

The conceptual framework model draws on constructs from all the previous models, as outlined in Table 6.1. These elements represent the core variables identified in earlier research while considering the additional constructs of organisational culture and the material information management system. The constructs of the CEBM for construction organisations were identified by reviewing the literature.

6.3 Model Structural Component

The study proposes a conceptual framework suggesting that the CEBM for construction organisations is influenced by several key factors: value proposition (VP), client segments (CS), channels (CH), client relationships (CR), key resources (KR), key activities (KA), key partnerships (KP), cost structure (CS), revenue streams (RS), take-back systems (TBS), material information management systems (MIMS), and organisational culture features (OCF). According to the framework, CEBM is a multi-faceted framework composed of these elements. This hypothesis with a schematic representation of these categories is provided in Fig. 6.1, based on the classifications outlined in Table 6.1. The theoretical basis for this framework is drawn from Lewandowski's [21] work, which was previously referenced.

The concept of a CEBM is closely rooted in a structured framework that considers various factors. Discussing the value proposition (the key element of focus) is difficult without acknowledging the influence of other exogenous factors. The CEBM reflects

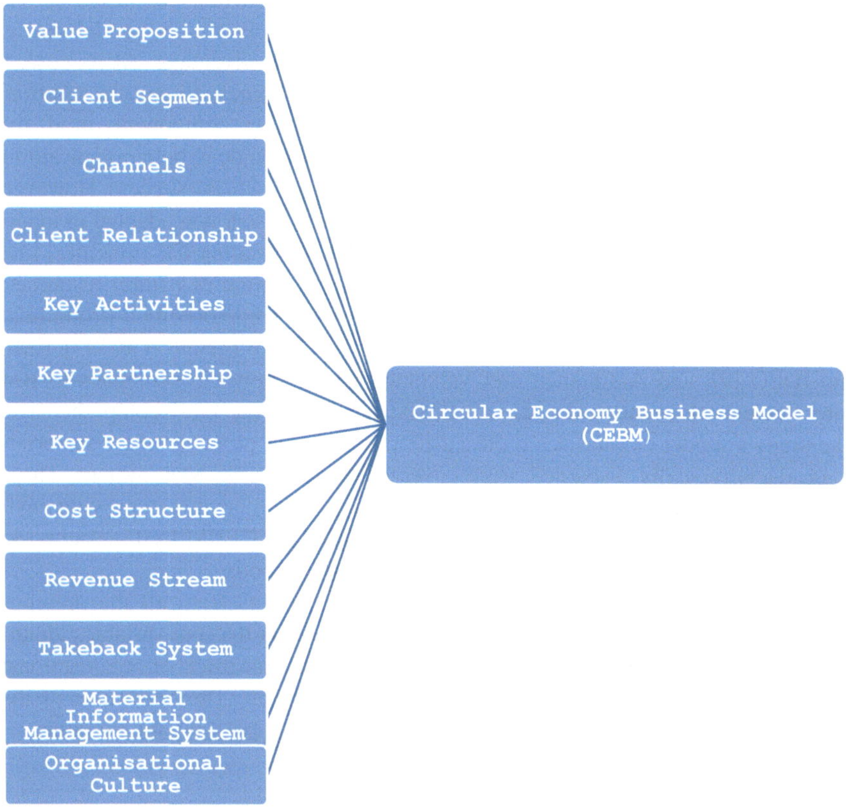

Fig. 6.1 Conceptual CE business model

both market orientation and business ventures. It is important to note that CEBMs are not fixed; they evolve to adapt to changing business environments. As such, entrepreneurs across different construction industry sectors can leverage this dynamic framework to create, deliver, and capture value for their clients.

6.4 Summary

This chapter presents a conceptual framework designed for adopting CEBM in construction organisations. The multidimensional framework consists of twelve key constructs: value proposition, client segment, channels, client relationship, key resources, key activities, key partnerships, cost structure, revenue streams, take-back system, material information management system, and organisational culture. These variables were derived from an extensive literature review conducted for this study. In addition, the chapter thoroughly explains the theoretical foundation behind the development of this conceptual framework for adopting CEBM in construction organisations.

References

1. Anderson JC, Narus JA, Van Rossum W (2006) Customer value propositions in business markets. Harv Bus Rev 84(3):90
2. Antikainen M, Valkokari K (2016, March) Framework for sustainable circular business model innovation. In: ISPIM innovation symposium. The International Society for Professional Innovation Management (ISPIM), p 1
3. Bakker CA, Den Hollander MC, Van Hinte E, Zijlstra Y (2014) Products that last: product design for circular business models. TU Delft Library
4. Barquet APB, de Oliveira MG, Amigo CR, Cunha VP, Rozenfeld H (2013) Employing the business model concept to support the adoption of product–service systems (PSS). Ind Mark Manage 42(5):693–704
5. Berning A, Venter C (2015) Sustainable supply chain engagement in a retail environment. Sustainability 7(5):6246–6263
6. De Jong E, Engelaer F, Mendoza M (2015) Realizing opportunities of a circular business model. http://circulatenews.org/2015/04/de-lage-landen-realising-theopportunities-of-a-circular-business-model
7. El-Haggar S (2007) Cleaner production. In: Sustainable industrial design and waste management: cradle-to-cradle for sustainable development. pp 21–84
8. Ellen MacArthur Foundation (2015) Growth within: A circular economy vision for a competitive Europe, vol 100. Ellen MacArthur Foundation, pp 81–92
9. Ellen MacArthur Foundation, & Ide (2017) Business model canva. http://circulardesignguide.com/post/circular-business-model-canvas
10. Gale R (2006) Environmental management accounting as a reflexive modernization strategy in cleaner production. J Clean Prod 14(14):1228–1236. https://doi.org/10.1016/j.jclepro.2005.08.008
11. Govindan K, Soleimani H, Kannan D (2015) Reverse logistics and closed-loop supply chain: a comprehensive review to explore the future. Eur J Oper Res 240(3):603–626

12. Jasch C (2006) Environmental management accounting (EMA) as the next step in the evolution of management accounting. J Clean Prod 14(14):1190–1193
13. Jasch C (2006) How to perform an environmental management cost assessment in one day. J Clean Prod 14(14):1194–1213
14. Joustra DJ, de Jong E, Engelaer F (2013) Guided choices: towards a circular business model. C2C BIZZ
15. Lacy DL (2022) Employee engagement and Learning for the Transformational Leader. In: Interdisciplinary and Practical Approaches to Managerial Education and Training. IGI Global Scientific Publishing, pp 133–146
16. Lacy P, Keeble J, McNamara R, Rutqvist J, Haglund T, Cui M, Cooper A, Pettersson C, Kevin E, Buddemeier P (2014) Circular advantage: innovative business models and technologies to create value in a world without limits to growth. Accenture
17. Lacy P, Rutqvist J (2015) Waste to wealth. Palgrave Macmillan UK.https://doi.org/10.1057/9781137530707
18. Lacy P, Long J, Spindler W, Lacy P, Long J, Spindler W (2020) The circular business models. In: The circular economy handbook: realizing the circular advantage. pp 17–42
19. Ladd T (2018) Does the business model canvas drive venture success? J Res Mark Entrep 20(1):57–69
20. Laubscher M, Marinelli T (2014) Integration of circular economy in business. In: Proceedings of the conference: going Green—care innovation. https://doi.org/10.13140/2.1.4864.4164
21. Lewandowski M (2016) Designing the business models for circular economy—towards the conceptual framework. Sustainability 8(1):43
22. Linder M, Williander M (2017) Circular business model innovation: inherent uncertainties. Bus Strateg Environ 26(2):182–196
23. Lüdeke-Freund F, Carroux S, Joyce A, Massa L, Breuer H (2018) The sustainable business model pattern taxonomy—45 patterns to support sustainability-oriented business model innovation. Sustain Prod Consum 15:145–162
24. Lüdeke-Freund F, Gold S, Bocken NM (2019) A review and typology of circular economy business model patterns. J Ind Ecol 23(1):36–61
25. Mentink B (2014) Circular business model innovation: a process framework and a tool for business model innovation in a circular economy. http://resolver.tudelft.nl/uuid:c2554c91-8aaf-4fdd-91b7-4ca08e8ea621
26. Osterwalder A, Pigneur Y (2010) Business model generation: a handbook for visionaries, game changers, and challengers, vol 1. Wiley
27. Payne A, Frow P (2017) Relationship marketing: looking backwards towards the future. J Serv Mark 31(1):11–15
28. Pearce II JA (2009) The profit-making allure of product reconstruction. MIT Sloan Manag Rev
29. Planing P (2015) Business model innovation in a circular economy reasons for non-acceptance of circular business models. Open J Bus Model Innovation 1(11):1–11
30. Porter ME (1980) Industry structure and competitive strategy: keys to profitability. Financ Anal J 36(4):30–41
31. Rifkin J (2014) The zero marginal cost society: the internet of things, the collaborative commons, and the eclipse of capitalism. Macmillan
32. Scott JT (2017) The sustainable business: a practitioner's guide to achieving long-term profitability and competitiveness. Routledge
33. Sheu JB (2014) Green supply chain collaboration for fashionable consumer electronics products under third-party power intervention—a resource dependence perspective. Sustainability 6(5):2832–2875
34. Subramanian N, Gunasekaran A (2015) Cleaner supply-chain management practices for twenty-first-century organizational competitiveness: practice-performance framework and research propositions. Int J Prod Econ 164:216–233
35. Teece DJ (2010) Business models, business strategy and innovation. Long Range Plan 43(2–3):172–194
36. Tukker A (2004) Eight types of product–service system: eight ways to sustainability? Experiences from SusProNet. Bus Strateg Environ 13(4):246–260

37. Van Ostaeyen J, Van Horenbeek A, Pintelon L, Duflou JR (2013) A refined typology of product–service systems based on functional hierarchy modeling. J Clean Prod 51:261–276
38. Van Renswoude K, Wolde AT, Joustra DJ (2015) Circular business models. Part 1: an introduction to IMSA's circular business model scan. https://groenomstilling.erhvervsstyrelsen.dk/sites/default/files/media/imsa_circular_business_models_-_april_2015_-_part_1.pdf
39. Webster FE (2002) Market-driven management: how to define, develop and deliver client value, 2nd edn. Wiley
40. Waste Resources Action Plan (WRAP) (2022) Waste avoidance and resource recovery strategy 2030. https://www.wasteauthority.wa.gov.au/images/resources/files/2022/06/Waste_Strategy_Action_Plan_2022-23.pdf
41. Zott C, Amit R, Massa L (2011) The business model: recent developments and future research. J Manag 37(4):1019–1042

Part III
Case Study

Chapter 7
Conceptual Framework Validation: A Delphi Approach

Abstract This study employed a Delphi approach to explore the key features and critical factors influencing the adoption of the Circular Economy Business Model (CEBM) in construction organisations; by gathering insights from a panel of experts, the research aimed to identify essential attributes for successfully implementing CEBM and forecast its future adoption within the sector. The Delphi process consisted of two rounds, enabling experts to reach a consensus on issues related to CEBM's application in construction. The study also examined the role of organisational functions in facilitating the transition to a circular economy, focusing on critical factors that may affect its adoption. The research highlighted a gap in existing literature regarding frameworks for implementing CEBM in construction, prompting the development of a comprehensive framework to guide this transition. A carefully selected expert panel was instrumental in validating the model and identifying primary and secondary attributes essential for CEBM adoption. This study contributes to understanding how construction organisations can effectively transition from linear practices to more sustainable, circular approaches.

7.1 Introduction

This book utilised a Delphi study to gather insights from experts on the importance and scope of various features of the circular economy business model (CEBM) relevant to construction organisations. This approach helped identify the key features necessary to implement CEBM in the construction sector successfully. In addition, expert opinions were analysed to determine the critical factors influencing the adoption of circular economy principles and assess their impact. Furthermore, to forecast the future adoption of CEBM, the study also explored the role of various organisational functions in driving this transition. The Delphi process consisted of two rounds, allowing experts to reach a consensus on the issues addressed in the questionnaire. The chapter provides an overview of the Delphi study methodology and details the composition of the expert panel.

© The Author(s), under exclusive license to Springer Nature Switzerland AG 2025

O. Otasowie et al., *Circular Economy Business Model for Construction Organisations*, Green Energy and Technology, https://doi.org/10.1007/978-3-031-88322-4_7

7.2 Delphi Study Methodology

The Delphi technique is structured to conduct in-depth assessment and discussion on a specific topic. According to Buckley [10], its origins trace back to the 1950s, when it was developed at the Rand Corporation as a tool for problem-solving and forecasting. The method comes from the ancient Greek temple at Delphi, which housed the famous oracle. In Greek mythology, this oracle was sought for guidance in predicting the outcomes of critical decisions, such as initiating a war. Norman Dalkey and Olaf Helmer pioneered the Delphi technique at the Rand Corporation. Initially, it was employed to analyse and predict trends in various domains, including weapon systems, space exploration, automation, population management, and scientific advancements, all with the broader aim of addressing challenges like war prevention [29]. Following its initial success, the approach was adopted in subsequent studies, becoming a strategic tool for the U.S. Department of Defense, government agencies, healthcare organisations, businesses, and industries to anticipate future developments and guide planning efforts. Turoff and Hiltz [46] noted that the work of Dalkey and Helmer was humorously dubbed "Delphi research" by their colleagues at the Rand Corporation. This nickname arose from the methodology's reliance on anonymity to gather expert opinions and forecast trends, particularly in nuclear science.

Falzarano and Zipp [16] describe the Delphi technique as a qualitative approach to achieve consensus among experts on a specific topic under investigation. It involves a structured communication process that facilitates expert agreement on a subject of interest [11]. The method operates through iterative rounds, during which experts provide responses and opinions, gradually working toward a consensus [20]. Ideally, the experts selected for a Delphi study should have specialised knowledge in the relevant field. This technique combines qualitative and quantitative outputs and is characterised by its normative, predictive, and exploratory features [13]. The process typically consists of multiple rounds, where the feedback from each round is shared with the experts. As a result, each expert's subsequent input is informed by the collective feedback from their peers. This iterative process allows coordinated group communication among geographically dispersed experts [35]. In a Delphi study, the researcher acts as a central point for gathering and analysing individual contributions from experts. After analysing the responses from each round, the researcher identifies patterns, including central and extreme tendencies, and validates the findings [20]. The feedback is then shared with the experts for further input, and this cycle continues until a consensus is reached on the subject under review.

While the Delphi technique is primarily considered a qualitative methodology [22], it has gained popularity for its ability to incorporate quantitative approaches by combining qualitative and quantitative methods [20]. A significant advantage of this technique is its ability to minimise biases often present in face-to-face expert meetings. The anonymity of the process ensures that experts do not know each other's identities, reducing the risk of individuals being influenced or reluctant to share their genuine opinions due to social pressures or fear of losing credibility [43].

Moreover, the Delphi method is beneficial for long-term forecasting, as it relies on expert opinions to generate insights about future trends and issues. Its reliance on structured questionnaires as the primary means of interaction ensures a systematic and unbiased approach to gathering expert insights.

Nevertheless, the Delphi technique has faced both criticism and support over time. A major critique is that it lacks scientific rigour, particularly concerning psychometric validity [41] and accuracy [7]. In addition, Ewing [15] argued that the Delphi method is often a last resort for tackling highly complex problems without established models. Similarly, Fletcher and Marchildon [17] observed that the technique's reliance on intuitive judgment is a core characteristic rather than a temporary measure. Steurer [43] highlighted three significant drawbacks of the Delphi method, and they include the low reliability of expert judgments, which makes the results heavily dependent on the selection of experts; the method's vulnerability to ambiguities in the questionnaires used for data collection in successive rounds; and the challenges in evaluating the expertise incorporated into the forecasts. In a separate analysis, Haynes and Shelton [23] distinguished between the execution and application of the Delphi technique. While the study acknowledged that some studies have applied the method poorly, they emphasised that this does not reflect a flaw in the method itself.

On the other hand, various studies advocate for the Delphi technique's utility. Chang and Ding [12] argued that its value lies not in generating public knowledge or individual data but in driving thoughtful judgment and collective wisdom. Kauko and Palmroos [30] identified its effectiveness in forecasting financial markets. Similarly, Helmer [24] noted the method's strength in working with "soft law" and "soft data." The study further argued that criticism based on its reliance on expert opinion and the absence of random sampling in expert selection misinterprets the technique's purpose. The study clarified that the Delphi method should not be equated with an opinion poll, as its goal is to achieve consensus rather than a majority opinion. Despite its limitations, Brill et al. [9] described the Delphi method as a powerful tool for achieving expert consensus on specific issues, especially when the experts are geographically dispersed and the information required is subjective. The study also highlighted its validation across diverse fields, including construction [4], commerce [1], journalism [12], healthcare [48], social sciences [33], and education [49].

7.2.1 Delphi-Specific Objectives

The review of existing literature highlighted various dimensions and attributes associated with CEBM adoption across different domains of human activity. Over time, numerous theories and models have been proposed to evaluate the readiness and effectiveness of the CEBM. Despite thoroughly examining these theories and the frameworks underpinning CEBM, there is a notable gap in research specifically addressing how CEBM adoption can be realised within construction organisations. It is worth emphasising that CEBMs offer construction organisations strategies to transition from traditional linear practices to a more sustainable, circular approach. Consequently, developing and implementing such a framework will likely face challenges, particularly in many countries. Considering the significant role of the construction sector in achieving sustainable development goals, there is a pressing need to develop a comprehensive framework that can facilitate the adoption of CEBM in construction organisations. To address this, the Delphi technique was employed to validate and test the attributes identified in the literature as critical for promoting the adoption of CEBM in the construction industry. Insights from the Delphi panel of experts were instrumental in identifying the primary and secondary attributes deemed crucial for this adoption process. The overarching objective was to evaluate and refine the conceptual model proposed in Chap. 6, which was developed after an extensive review of theories and frameworks related to CEBM. In line with the aim of the Delphi study, developing a framework to guide CEBM adoption in construction organisations, the study established the following specific objectives to be achieved.

- **DSO1**: To determine the main attributes of CEBM framework for construction organisations.
- **DSO2**: To determine the sub-attributes of CEBM framework for construction organisations.
- **DSO3**: To predict the critical factors that could affect the adoption of a CEBM framework in construction organisations.
- **DSO4**: To determine the extent to which policies can drive the adoption of a CEBM framework in construction organisations.
- **DSO5**: To identify the extent to which the benefits of a CEBM framework can drive the adoption of a CEBM framework in construction organisations.

7.2.2 Selection of Panel Experts

The success of a Delphi study significantly relies on the careful and deliberate selection of the experts who form the panel [11]. These experts must be genuinely interested in the topic and hold a vested stake in the subject matter. Such attributes enhance their commitment to the process and ensure that their expertise meaningfully contributes to the study. Hallowell [19] highlights that experts involved in the Delphi process are typically professionals or researchers with considerable experience and knowledge, demonstrated through various qualifications such as work history, academic achievements, professional roles, and published works. However, the concept of "expert" remains subject to debate. Hasson et al. [22] point out disagreements over classifying a professional as an "expert." For this study, McKenna [34] definition of an expert is adopted to provide clarity, identifying an expert panel composed of informed individuals. Similarly, Goodman [18] suggests that the Delphi method avoids random sampling in favour of selecting experts or well-informed advocates. Helmer [24] further argues that since a Delphi study is not an opinion poll, relying on random sampling from the expert population is suboptimal. Instead, the focus should be on applying predetermined criteria to select experts and facilitating effective communication while maintaining anonymity.

Thus, a significant challenge in forming a Delphi panel is the risk of selecting individuals with similar perspectives or close associations, which could undermine the process's robustness. To counter this, Hemmat et al. [25] advocate involving key experts in the relevant field to incorporate the latest insights and inform policymaking. They also emphasise integrating these expert opinions with existing knowledge to create a comprehensive research foundation. Nevertheless, bias in panel selection remains a concern, as experts may bring preconceived notions to the process. Hasson et al. [22] stress the importance of neutrality and impartiality during the selection phase, though the study acknowledges that misconceptions may not be entirely avoidable. Clear selection criteria should be established to evaluate outcomes and ensure the study's relevance to other contexts and populations [28].

Niederberger and Spranger [36] propose that several factors should guide the selection of panel experts. Key recommendations include ensuring experts possess deep knowledge of the topic and represent their professional field to facilitate broader acceptance of their recommendations. Similarly, Adler and Ziglio [2] outline four essential qualifications for panel experts in a Delphi study: expertise and knowledge of the subject, willingness and ability to participate, sufficient time to engage throughout the process, and strong communication skills.

For this book, experts for the Delphi panel were selected based on meeting a minimum of five specified criteria, which include:

1. *Knowledge*: Individuals must possess substantial expertise in information and communication technology, facilities management, and construction management.
2. *Academic Qualification*: Individuals should hold a degree in a construction-related discipline, such as a National Diploma, Bachelor's, Master's, or Ph.D.
3. *Experience*: A background of involvement in construction projects incorporating circular economy principles is essential. This experience may stem from roles within consultancy firms, contracting organisations, or government entities.
4. *Employment*: Individuals must hold or have previously held professional roles within institutions, agencies, departments, businesses, organisations, or companies.
5. *Authorship*: Individuals should have contributed to the academic community by authoring or co-authoring articles in peer-reviewed journals focused on circular economy or sustainable construction. They should also have presented their work at conferences, symposiums, professional meetings, or workshops.
6. *Recognition*: Individuals must have served as peer reviewers for at least one manuscript from the editor of a respected journal emphasising circular economy principles.
7. *Teaching*: Individuals must have experience teaching classes or delivering lessons focused on circular economy or sustainable construction. They should

also have made presentations at workshops aimed at providing professional, advisory, or technical expertise in these areas.
8. *Research*: Individuals must have participated in research activities, submitted at least one research proposal, or secured a research grant or funding from a governmental or private funding agency in the circular economy or sustainable construction field.
9. *Professional Membership*: Membership in a recognised professional organisation related to construction is required.
10. *Willingness*: Individuals must demonstrate a strong commitment and readiness to engage actively in the Delphi process.

An expert panel was recruited via email, which included a brief overview of the study's objectives. Consent was requested for participation in the Delphi study, and those who agreed were provided with a detailed explanation of the Delphi process. In addition, confirmed participants were asked to submit their curriculum vitae (CV) to verify their expertise and meet the specified selection criteria. After reviewing the CVs, all selected experts were confirmed to meet the eligibility requirements. A total of 30 invitations were sent, resulting in 11 responses.

Subsequently, the initial round of the questionnaire, including closed and open-ended questions, was distributed to the 11 experts. These individuals were retained for the Delphi process's first and second rounds. This number was considered sufficient, as prior research indicates no strong relationship between the size of a Delphi panel and the method's efficiency or accuracy [8]. Following these recommendations and acknowledging that the success of the Delphi method is more reliant on reaching a consensus than on statistical power, a panel of 11 experts was deemed appropriate for this study.

7.2.3 Determining the Panel Size

Vernon [47] highlighted the absence of a universally agreed-upon standard for determining the panel size in Delphi studies, with no clear guidelines to define what constitutes a "small" or "large" sample. There is significant variability in expert panel sizes and the criteria used to justify these choices. Studies employing the Delphi technique have used widely varying sample sizes, as Reid [40] noted, with published articles reporting panel sizes ranging from 10 to 1685 participants across different fields. This variation underscores the lack of consensus among researchers regarding an ideal sample size for Delphi studies. For instance, Strasser et al. [44] utilised six experts to study drug interactions in ambulatory pharmacies. Similarly, Harder et al. [21] recommended a panel of 12 participants for Delphi studies, while Howell and Kemp [27] used 13 experts to explore the diverse skill sets of young children. In a study proposing recommendations for treating gastroesophageal diseases, Armstrong et al. [6] engaged a panel of 23 experts. Dunn [14] suggested that 10–30 experts might be appropriate, advocating for larger sample sizes to capture a broader range of opinions when addressing complex issues. Dunn [14] further recommended including participants from diverse groups with varying degrees of influence, interest, and affiliations to ensure comprehensive representation, an approach reflected in the foundation of the current study.

Determining sample size in Delphi studies often depends on the research context and available resources. Many studies rely on convenience sampling, driven by expert availability and logistical constraints. Defining a minimum panel size is challenging due to the absence of a universally accepted standard. Andranovich [3] suggested that a homogeneous group of experts might require only 10–15 participants, whereas more diverse groups should include more participants [21]. Skulmoski et al. [42] identified several factors that researchers should consider when deciding on the sample size for a Delphi study, emphasising that these decisions must align with the study's goals and context. They include:

1. *Homogeneity or Heterogeneity of the Sample*: When the participants in the group share similar characteristics (homogeneous), a sample size of 10–15 individuals is typically sufficient to achieve reliable results. In contrast, for groups with diverse characteristics (heterogeneous), a larger sample size is recommended to ensure the validity of the findings.
2. *Impact on Decision Quality*: Increasing the sample size generally improves the quality of decisions and reduces the likelihood of errors within the group. However, managing the Delphi process can become overly complicated when the sample size exceeds a specific limit.
3. *Verification and Credibility*: Larger groups enhance the credibility of outcomes in the Delphi process through stronger external and internal validation. Nonetheless, additional follow-up studies can still confirm outcomes from smaller groups. A smaller group was utilised for this study, and the results will be validated using a questionnaire survey.

7.3 The Delphi Process

7.3.1 Delphi Iterations

The Delphi process is a structured method that involves designing questionnaires, analysing responses, and incorporating participant feedback over multiple rounds until a consensus is achieved. According to Suris and Akre [45], this process entails distributing and retrieving sequential rounds of questionnaires to gather iterative feedback on a particular subject. However, no universally agreed-upon number of rounds for a Delphi study exists. Kloser [32] recommended three rounds, whereas Kim et al. [31] suggested two to ten rounds, emphasising that more rounds can enhance accuracy. Nonetheless, Powell [38] highlighted that increasing the number of rounds may lead to higher time demands, participant fatigue, and attrition rates.

This study sought to achieve consensus after two rounds of the Delphi process, focusing on identifying attributes for CEBM in construction organisations. A comprehensive and detailed questionnaire was developed to address issues related to the framework of CEBM for construction organisations. The questionnaire was informed by an extensive review of relevant literature, theories, and existing frameworks and was sent to the selected Delphi panel members via email. The Delphi process in this study involved two iterative rounds, each taking approximately one month to complete. Experts were given ample time to provide thoughtful contributions. The first round's questionnaire was based on findings from the literature review and included closed and open-ended questions, allowing participants to introduce new ideas. The feedback from the first round informed the development of the second-round questionnaire, which focused on refining and evaluating the proposed attributes.

In the second round, closed-ended questions were used to gauge the experts' agreement on the attributes, including those newly proposed based on first-round feedback. The responses from both rounds were thoroughly analysed to determine the level of consensus among the participants. By the end of the second round, consensus was reached on nearly all attributes and related issues under consideration. The findings from the Delphi study validated the conceptual framework of this study. This validated framework was further assessed through a follow-up questionnaire survey, ensuring the robustness and reliability of the results.

7.3.2 Achieving Consensus in the Delphi Study

A primary reason for conducting a Delphi study is to establish consensus by utilising the insights of a panel of experts on a particular topic. In this study, reaching an agreement on the identified primary attributes and their sub-attributes was crucial. To achieve this, the study relied on the group median and interquartile deviation (IQD) to assess consensus. Specifically, consensus was considered reached when the deviation in the group median of responses for a given question did not exceed one unit. The IQD was also applied to individual questions, calculated as the absolute difference between the 75th (third quartile) and 25th (first quartile) percentiles. These percentiles represent values in a data distribution, where Q1 (25th percentile) marks the lower quartile, Q2 (50th percentile) represents the median, and Q3 (75th percentile) marks the upper quartile. A smaller IQD value indicates more substantial agreement, while a larger IQD reflects more significant disagreement.

Determining consensus in a Delphi study is inherently complex, as there are no universally accepted guidelines for achieving it. Consensus generally implies agreement and can be established through iterative evaluations, resulting in converging expert opinions or stability in responses over successive rounds [31]. Researchers have employed various statistical methods to measure consensus in Delphi studies. For instance, Rayens and Hahn [39] used means and standard deviations, where a reduction in standard deviation across rounds signified increased consensus.

McKenna [34] evaluated frequency distributions with a criterion of 51% agreement. Similarly, IQD has been widely utilised for consensus determination [5]. According to Olij et al. [37], an IQD value of 1.00 or less indicates consensus, while values exceeding 1.00 suggest continued divergence in opinions until eventual agreement is achieved [5]. Holey et al. [26] also outlined specific criteria for measuring consensus in Delphi studies to include:

1. The percentage of responses.
2. For each question, the percentage reflects the level of agreement across different response rates.
3. A detailed analysis of each question, including its median, standard deviation, and corresponding group rankings.
4. An analysis of each question's mean values, standard deviations, and group rankings.
5. An evaluation of the weighted kappa (k) values to compare the level of agreement across rounds and identify any eliminated chance agreement.

In addition, Holey et al. [26] outlined the following criteria to demonstrate when consensus has been reached.

1. An increase in the percentage of agreement.
2. Greater alignment in the rankings of importance.
3. Higher kappa values observed.
4. A decrease in standard deviation measurements.
5. A smaller range of responses.
6. A decline in the number of comments provided across successive rounds.

As previously mentioned, no universally accepted standards for determining consensus in a Delphi process exist. However, a shared principle across various proposed approaches is the necessity for a certain degree of alignment in the experts' opinions. This alignment is often reflected in their convergence around a central tendency based on subjective measures. In this thesis, the following criteria were established to define consensus within the Delphi study:

1. Approximately 60% of the responses to the given questions fall into either positive or negative categories.
2. The absolute deviation for the provided questions does not exceed one unit.
3. An IQD (Interquartile Deviation) value below 1.00 indicates a high level of agreement, with IQD $= 1$ reflecting a very high consensus.

Therefore, this study utilised the following scales to categorise consensus:

1. Strong consensus—median 9–10, mean 8–10, IQD ≤ 1 and $\geq 80\%$ (8–10).
2. Good consensus—median 7–8.99, mean 6–7.99, IQD $\geq 1.1 \leq 2$ and $\geq 60\% \leq 79\%$ (6–7.99).
3. Weak consensus—median ≤ 6.99, mean ≤ 5.99, IQD $\geq 2.1 \leq 3$ and $\leq 59\%$ (5.99).

7.4 Summary

The Delphi approach used for this study was provided in this chapter. It also presented the Delphi-specific objectives, selection of panel experts, and establishing the panel size. The chapter also discussed the Delphi process, which includes Delphi iterations and achieving consensus in the Delphi study. The data presentations and discussions from the Delphi study are provided in the next chapter.

References

1. Addison T (2003) E-commerce project development risks: evidence from a Delphi survey. Int J Inf Manage 23(1):25–40
2. Adler M, Ziglio E (1996) Gazing into the oracle: the Delphi method and its application to social policy and public health. Jessica Kingsley Publishers, London
3. Andranovich G (1995) Developing community participation and consensus: the Delphi technique. California State University, Los Angeles, CA
4. Aghimien D, Aigbavboa C, Oke A (2020) Critical success factors for digital partnering of construction organisations: a Delphi study. Eng Constr Archit Manag 27(10):3171–3188
5. Aigbavboa C (2013) An integrated beneficiary centred satisfaction model for publicly funded housing schemes in South Africa. PhD thesis. University of Johannesburg
6. Armstrong D, Marshall JK, Chiba N, Enns R, Fallone CA, Fass R, Hollingworth R, Hunt RH, Kahrilas PJ, Mayrand S, Moayyedi P, Paterson WG, Sadowski D, van Zanten SJ (2005) Canadian consensus conference on the management of gastroesophageal reflux disease in adults–update 2004. Can J Gastroenterol Hepatol 19(1):15–35
7. Avella J (2016) Delphi panels: research design, procedures, advantages, and challenges. Int J Dr Stud 11:305
8. Balasubramanian R, Agarwal D (2012) Delphi technique: a review. J Public Health Dent 3(2):16–26
9. Brill J, Bishop M, Walker A (2006) The competencies and characteristics required of an effective project manager: a web-based Delphi study. Educ Tech Res Dev 54(2):115–140
10. Buckley C (1995) Delphi: a methodology for preferences more than predictions. Libr Manag 16(7):16–19
11. Chan APC, Yung EHK, Lam PTI, Tam CM, Cheung SO (2001) Application of Delphi method in selection of procurement systems for construction projects. Constr Manage Econ 19(7):699–718
12. Chang J, Ding R (2022) Theories of journalism in the digital era: knowledge, value, and conceptual framework. In: Digital journalism in China. Routledge, New York, pp 8–22
13. Cuhls K (2003) Delphi method. Available at: http://www.unido.org/fileadmin/import/16959_DelphiMethod.pdf. Accessed 18 Nov 2024
14. Dunn WN (1994) Public policy analysis: an introduction. Prentice-Hall, Englewood Cliffs, NJ
15. Ewing RP (1979) The uses of futurist techniques in issues management. Public Relations Q 4:15–18
16. Falzarano M, Zipp G (2013) Seeking consensus through the use of the Delphi technique in health sciences research. J Allied Health 42(2):99–105
17. Fletcher A, Marchildon G (2014) Using the Delphi method for qualitative, participatory action research in health leadership. Int J Qual Methods 13(1):1–18
18. Goodman CM (1987) The Delphi technique: a critique. J Adv Nurs 12:729–734
19. Hallowell M (2008) A formal model of construction safety and health risk management. PhD dissertation. Oregon State University

20. Hallowell MR, Gambatese JA (2010) Qualitative research: application of the Delphi method to CEM research. J Constr Eng Manag 136(1):99–107
21. Harder A, Place N, Scheer S (2010) Towards a competency-based extension education curriculum: a Delphi study. J Agric Educ 51(3):44–52
22. Hasson F, Keeney S, McKenna H (2000) Research guidelines for the Delphi survey technique. J Adv Nurs 32(4):1008–1015
23. Haynes C, Shelton K (2018) Delphi method in a digital age: practical considerations for online Delphi studies. In: Wang, V, Reid T (eds) Handbook of research on innovative techniques, trends, and analysis for optimized research methods. Information Science Reference, Hershey, PA, pp 132–151
24. Helmer O (1977) Problems in futures research: Delphi and causal cross-impact analysis. Futures 9(1):17–31
25. Hemmat M, Ayatollahi H, Maleki M, Saghafi F (2021) Health information technology foresight for Iran: a Delphi study of experts' views to inform future policymaking. Health Inf Manage J 50(1–2):76–87
26. Holey EA, Feeley JL, Dixon J, Whittaker VJ (2007) An exploration of the use of simple statistics to measure consensus and stability in Delphi studies. BMC Med Res Methodol 7(52):1–10
27. Howell S, Kemp C (2005) Defining early number sense: a participatory Australian study. J Educ Psychol 25(5):555–571
28. Iqbal S, Pipon-Young L (2009) The Delphi method. Nurs Res 46(2):116–118
29. Jones T (1980) Options for the future: a comparative analysis of policy-oriented forecasts. Praeger, New York
30. Kauko K, Palmroos P (2014) The Delphi method in forecasting financial markets: an experimental study. Int J Forecast 32(2):313–327
31. Kim J, King B, Kim S (2022) Developing a slow city tourism evaluation index: a Delphi-AHP review of Cittaslow requirements. J Sustain Tour 30(4):846–874
32. Kloser M (2014) Identifying a core set of science teaching practices: a Delphi expert panel approach. J Res Sci Teach 51(9):1185–1217
33. Landeta J (2006) Current validity of the Delphi method in social sciences. Technol Forecast Soc Chang 73(5):467–482
34. McKenna H (1994) The Delphi technique: a worthwhile research approach for nursing? J Adv Nurs 19(6):1221–1225
35. McMillan S, King M, Tully M (2016) How to use the nominal group and Delphi techniques. Int J Clin Pharm 38(3):655–662
36. Niederberger M, Spranger J (2020) Delphi technique in health sciences: a map. Front Public Health 8:457
37. Olij B, Erasmus V, Kuiper J, van Zoest F, Van Beeck E, Polinder S (2017) Falls prevention activities among community-dwelling elderly in the Netherlands: a Delphi study. Injury 48(9):2017–2021
38. Powell C (2003) The Delphi technique: myths and realities. Methodol Issues Nurs Res 41:376–382
39. Rayens MK, Hahn EJ (2000) Building consensus using the policy Delphi method. Policy, Polit, Nurs Pract 1(2):308–315
40. Reid NG (1988) The Delphi technique: its contribution to the evaluation of professional practice. In: Ellis R (ed) Professional competence and quality assurance in the caring professions. Chapman and Hall, New York
41. Sackman H (1974) Delphi assessment: expert opinion, forecasting and group process. Rand Corporation, Santa Monica, CA
42. Skulmoski GJ, Hartman FT, Krahn J (2007) The Delphi method for graduate research. J Inf Technol Educ 6:1–11
43. Steurer J (2011) The Delphi method: an efficient procedure to generate knowledge. Skeletal Radiol 40(8):959–961
44. Strasser S, London L, Kortenbout E (2005) Developing a competence framework and evaluation tool for primary care nurses in South Africa. Educ Health 18(2):133–144

45. Suris J, Akre C (2015) Key elements for, and indicators of, a successful transition: an international Delphi study. J Adolesc Health 56(6):612–618
46. Turoff M, Hiltz S (1996) Computer-based Delphi processes. In: Adler M, Ziglio E (eds) Gazing into the oracle: the Delphi technique and its application to social policy and public health. Jessica Kingsley, London
47. Vernon W (2009) The Delphi technique: a review. Int J Ther Rehabil 16(2):69–76
48. Whitman NI (1990) The committee meeting alternative: using the Delphi technique. J Nurs Adm 51(1):57–68
49. Yousuf MI (2007) The Delphi technique. Essays Educ 20:80–89

Chapter 8
Results of the Delphi Study

Abstract This study investigates the key attributes that define the circular economy business model (CEBM) for construction organisations through a Delphi methodology. In two rounds, the study engaged a panel of experts to identify and evaluate the main and sub-attributes shaping CEBM within the construction sector. The Delphi process facilitated consensus-building among experts on critical factors influencing the integration of CE principles into business models in this industry. The findings highlight the significance of specific attributes and constructs within the CEBM framework, emphasising their roles in driving sustainable practices in construction organisations. The study also examines the extent to which policies and other driving factors can influence the adoption of CEBM frameworks. Comprehensive analyses of the experts' responses revealed a shared understanding of the essential components of a robust CEBM framework, offering valuable insights into its development and implementation. Ultimately, the study provides a foundational framework for guiding the adoption of CEBM in construction organisations, offering practical implications for policymakers and industry stakeholders in advancing sustainable construction practices.

8.1 Introduction

The circular economy business model (CEBM) attributes on construction organisations were examined through a Delphi study. This study aimed to identify the key attributes that shape CEBM in the construction sector. The Delphi process consisted of two rounds, allowing experts to reach a consensus on the issues addressed in the questionnaire. Experts participating in the Delphi study also provided insights on critical factors affecting the integration of CE principles into business models within the industry. The study sought to achieve a consensus among experts on the identified attributes and variables related to CEBM. This section summarises the findings, detailing the influence of specific CE business model elements on the overall framework. The chapter concludes with a comprehensive discussion of the results aligned with the study's objectives.

© The Author(s), under exclusive license to Springer Nature Switzerland AG 2025

O. Otasowie et al., *Circular Economy Business Model for Construction Organisations*, Green Energy and Technology, https://doi.org/10.1007/978-3-031-88322-4_8

8.2 Background Information of Experts

Eleven (11) experts participated in both phases of the Delphi study. The selection of panel members prioritised individuals with expertise in construction management, particularly those with experience related to circular economy principles. This emphasis was crucial considering the study's focus on CEBM. In addition, to ensure a comprehensive perspective, the panel included a balanced mix of academics and industry practitioners. Table 8.1 provides an overview of the experts selected for the study, detailing how they met the established selection criteria. Regarding academic qualifications, six experts, representing 54.55% of the total, held a Master's degree, while five experts, accounting for 45.45%, had a Doctorate. The experts' areas of specialisation were diverse, with five focusing on Sustainable Construction Management, one on Sustainability and Construction Digitisation, one on Environmental Sustainability in Construction, one on Quantity Surveying and Sustainable Construction, one on Architecture and Sustainable Construction, one on Civil Engineering and Sustainable Energy, and one on Construction Management and Engineering with a focus on the Circular Economy. Regarding professional experience, two experts had 1–5 years of experience, six had 6–10 years, one had 11–15 years, and two had 16–20 years. Also, the employment affiliations of the experts showed that six were associated with contracting organisations, while government agencies employed five.

The Delphi questionnaire served as the primary tool for gathering expert insights. Each round of the Delphi process involved a tailored questionnaire, with the second round built upon the findings from the first. The questionnaire for the initial round was developed by identifying primary and sub-attributes from a comprehensive review of existing literature. These attributes were systematically combined to design the first-round questionnaire, a brainstorming tool to generate a detailed list of key features relevant to CEBM in construction organisations. This round included open-ended questions, allowing experts to propose additional features within the reviewed categories. The responses from the first round informed the development of the second-round questionnaire. Statistical measures such as the median and the Interquartile Deviation (IQD) were employed to determine the level of consensus among the experts regarding CEBM features for construction organisations.

The objective of the second round of the Delphi survey was to allow expert panel members to review, evaluate, and provide feedback on the features of CEBM for construction organisations, which had been initially proposed during the first round. This stage included a set of closed-ended questions aimed at gauging the experts' levels of agreement, disagreement, or justification regarding the proposed aspects of CEBM for construction organisations. Expert opinions largely aligned in this round, indicating broad consensus on the proposed features. As a result, a third round of the survey was deemed unnecessary. Statistical measures such as the mean, median, standard deviation, and interquartile deviation (IQD) were calculated for each question to assess the level of consensus. Experts were asked to explain their reason if responses deviated by two points or more from the median. Ultimately, the experts' responses

Table 8.1 The experts' background summary

Category	Classification	No. of experts	Percentage
Academic qualification	PhD	5	45.45
	Master's degree	6	54.55
	Total	**11**	**100**
Area of specialisation	Sustainable construction management	5	45.45
	Sustainability, construction digitisation	1	9.09
	Environmental sustainability in construction	1	9.09
	Quantity surveying, sustainable construction	1	9.09
	Architecture, sustainable construction	1	9.09
	Civil engineering, sustainable energy	1	9.09
	Construction management and engineering with specialisation in circular economy	1	9.09
	Total	**11**	**100**
Years of experience	1–5 years	2	18.18
	6–10 years	6	54.55
	11–15 years	1	9.09
	16–20 years	2	18.18
	Total	**11**	**100**
Employment	Government	5	45.45
	Contracting	6	54.55
	Total	**11**	**100**

showed convergence across the individual questions, demonstrating a successful consensus on the features of CEBM for construction organisations.

8.2.1 DSO1 to Determine the Main Attributes of CEBM for Construction Organisations

An extensive literature review identified key attributes characterising the CEBM for construction organisations. These attributes were subsequently presented to a panel of experts to assess their relevance in the framework of CEBM within the construction sector. Using an ordinal scale, the experts ranked each feature on a scale from 1 to

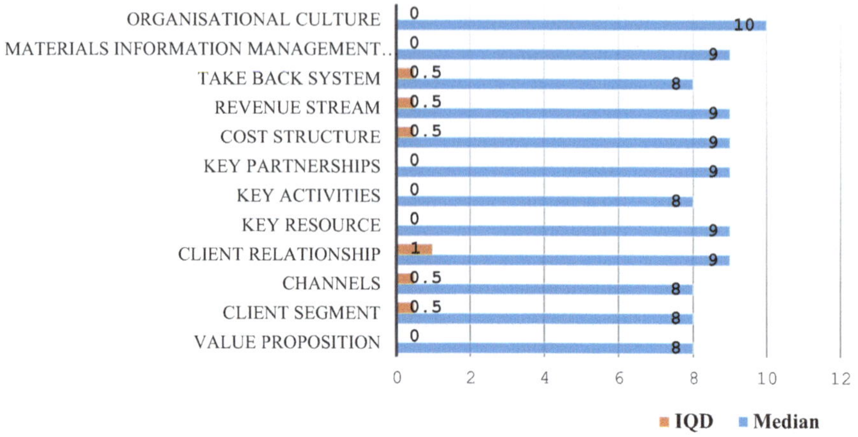

Fig. 8.1 Main attributes of CEBM for construction organisations

10, where 1 signified no significance and 10 denoted high significance. The analysis revealed twelve primary features of CEBM for construction organisations. Among these, seven features were rated as having very high significance (VHS) based on median scores between 9.0 and 10.0 (VHS: 9.00–10.00) (Fig. 8.1). The remaining five features were deemed to have high significance (HS), with median scores of 8.0 (HS: 7.00–8.99). Notably, no features fell within the medium significance (MS) range (MS: 4.00–6.99), and none were considered insignificant in determining the applicability of CEBM to construction organisations. Furthermore, the IQD values derived from the experts' evaluations demonstrated strong consensus across all features, with IQD scores meeting the study's threshold of less than or equal to 1.00. This consistency underscores the reliability of the expert assessments in prioritising the identified features.

8.2.2 DSO2 to Determine the Sub-attributes of CEBM for Construction Organisations

The literature review provided a thorough examination, leading to the identification of key features of CEBM for construction organisations. These features served as the foundational constructs for the proposed conceptual model. Each main feature included several sub-features that served as indicators to assess the underlying constructs. These sub-features were subsequently presented to an expert panel during the Delphi study to evaluate their significance within the context of the CEBM framework for construction organisations. The evaluation utilised an ordinal scale of 1–10, where 1 indicated very low significance, and 10 represented very high significance. The initial list of sub-features under the "value proposition" construct comprised 22 variables. As illustrated in Fig. 8.2, seven variables were rated as having very high significance (VHS: 9.00–10.00), while 15 were deemed to have high significance (HS: 7.00–8.99). Notably, none of the variables were considered insignificant to the CEBM framework for construction organisations.

Also, the IQD scores for 21 of the 22 variables ranged from 0.00 to 1.00, signifying a strong consensus among the experts. The only sub-attribute with above 1.00 IQD is "waste handling and processing". However, it is regarded as having high significance. This demonstrates consistent agreement regarding the significance of the "value proposition" construct, confirming its critical role in discussions on the CEBM framework for construction organisations.

For the client segment attribute, the sub-feature consisted of seven variables. Among these, three variables were identified as having very high significance (VHS: 9.00–10.00) in influencing CEBM for construction organisations. The remaining four variables were classified as having high significance (HS: 7.00–8.99). Notably, none of the variables were considered insignificant, as shown in Table 8.2. Also, based on the IQD values, all the variables achieved a consensus level.

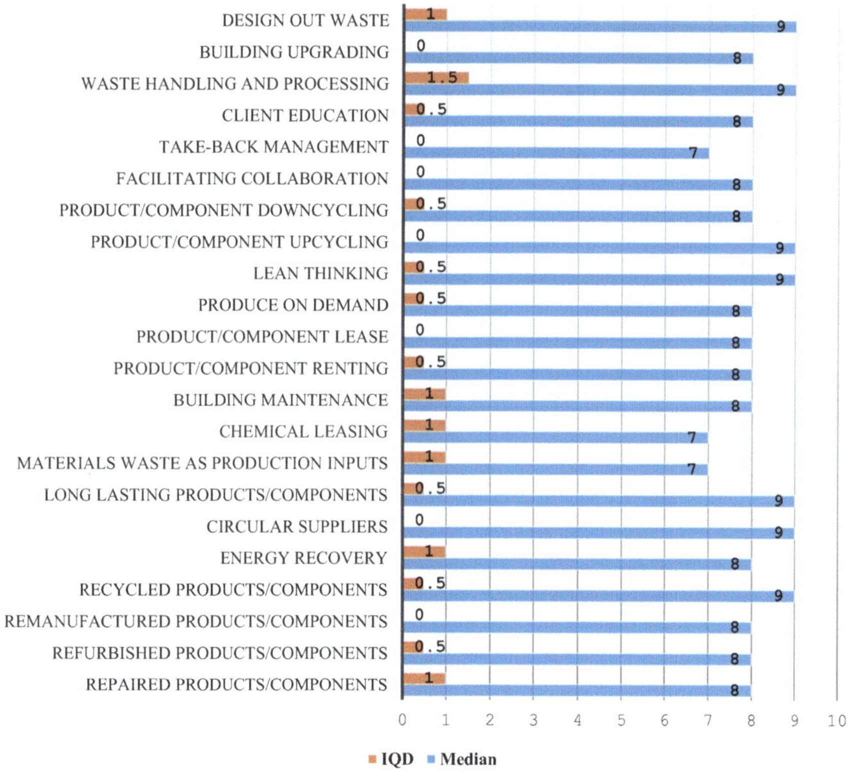

Fig. 8.2 Sub-attributes of value proposition of CEBM for construction organisations

Table 8.2 Client segment

Features	Median	Mean	SD	IQD
Quality conscious clients	9	8.73	0.47	0.50
Cost-conscious clients	9	8.64	0.92	0.00
Green clients	9	8.91	0.30	0.00
Business-to-business clients	8	7.91	0.70	0.50
Business to clients suppliers	8	7.91	0.70	0.50
Business-to-business suppliers	8	7.82	0.87	0.50
Clients to clients suppliers	8	7.82	0.87	0.50

Note IQD = interquartile deviation; SD = standard deviation

The third sub-feature, identified as channels, encompasses 13 variables. According to Table 8.3, expert evaluations within the framework of the CEBM for construction organisations determined that one variable held very high significance (VHS: 9.00–10.00). In addition, 11 variables were classified as highly significant (HS: 7.00–8.99), while one was rated as moderately significant (MS: 4.00–6.99). Notably, none of the variables were considered insignificant in influencing the framework of CEBM for construction organisations. Moreover, the IQD values for all variables fell within the range of 0.00–1.00, indicating consensus among the experts.

The fourth sub-feature, client relationship, comprises 13 variables. As shown in Table 8.4, the panel of experts evaluating the CEBM framework for construction organisations rated five of these variables as having very high significance (VHS: 9.00–10.00), while the remaining eight were classified as having high significance

Table 8.3 Channels

Features	Median	Mean	SD	IQD
Word of mouth	8	7.73	1.10	0.50
Direct personal selling	8	7.82	0.75	0.00
Wholesalers	8	7.73	0.47	0.50
Retailers	7	7.18	0.75	0.00
Websites	8	7.73	1.35	0.00
Web advertisements	9	8.55	1.57	0.00
Emails	8	7.27	1.49	1.00
Distributors	8	7.82	1.17	0.00
Social media	8	7.91	1.14	0.50
Video conferences	8	7.55	1.29	0.50
Newsletters	7	6.82	1.17	0.00
Directories/listing	6	6.18	1.25	0.50
Electronic-learning	8	7.82	0.60	0.50

Note IQD = interquartile deviation; SD = standard deviation

(HS: 7.00–8.99). Notably, none of the variables were considered insignificant within the context of the CEBM framework for construction organisations. Also, consensus was reached for all variables, as indicated by their IQD values.

The following sub-feature is key resources, encompassing seven variables. As detailed in Table 8.5, the panel of experts evaluating the CEBM framework for construction organisations rated four variables as having very high significance (VHS: 9.00–10.00), while the remaining three were classified as highly significant (HS: 7.99–8.00). Notably, none of the variables were deemed insignificant within this framework. Furthermore, all variables achieved consensus, as indicated by their IQD values.

The sixth sub-feature, key activities, encompasses 13 variables. As shown in Table 8.6, the panel of experts rated four of these variables as having very high significance (VHS: 9.00–10.00) within the CEBM framework for construction organisations. The remaining nine variables were categorised as highly significant (HS: 7.99–8.00). Notably, no variables were identified as insignificant in the CEBM framework. Furthermore, consensus was reached for all variables, as indicated by their IQD values.

The following sub-feature is key partnerships, encompassing thirteen variables. As detailed in Table 8.7, the panel of experts evaluating the CEBM framework for construction organisations rated eight variables as having very high significance

Table 8.4 Client relationship

Features	Median	Mean	SD	IQD
Client identification				
Target clients' analysis	8	8.09	0.54	0.00
Client segmentation	8	8.09	0.54	0.00
Client attraction				
Direct marketing	8	8.27	0.47	0.50
Web advertisements	8	7.91	0.83	0.00
Emails	7	7.09	0.54	0.00
Social media	9	8.82	0.40	0.00
Newsletters	8	7.91	0.83	0.00
Electronic learning	8	7.91	0.83	0.00
Client retention				
One-to-one marketing	9	8.91	0.30	0.00
Complaint management	9	8.91	0.54	0.00
Client development				
Client lifetime value	9	8.73	0.47	0.50
Upselling/Cross-selling	8	8.18	0.60	0.50
Market basket analysis	9	9.00	0.45	0.00

Note IQD = interquartile deviation; SD = standard deviation

Table 8.5 Key resources

Features	Median	Mean	SD	IQD
Circular sourcing	8	8.18	0.60	0.50
Substitution of resources with better-performing materials	9	8.82	0.40	0.00
Direct visualisation of materials (digitalisation)	8	8.18	0.40	0.00
Energy from renewable sources	9	9.00	0.00	0.00
Operating in efficient buildings	9	8.82	0.60	0.50
Sustainable production locations	9	9.00	0.45	0.00
Land restoration or reclamation	8	8.27	0.79	0.50

Note IQD = interquartile deviation; SD = standard deviation

Table 8.6 Key activities

Features	Median	Mean	SD	IQD
Increasing performance				
Housekeeping	8	7.73	0.90	1.00
Process control	8	8.18	0.60	0.50
Equipment modification and technology changes	9	9.00	0.45	0.00
Sharing and virtualisation—desk sharing, audio and video conferences	8	7.82	1.08	0.00
Product design				
Less raw material or energy	9	8.91	0.54	0.00
Prolonging product/component life	9	9.09	0.54	0.00
Resource life extension	9	8.91	0.54	0.00
Others				
Remanufacturing components	8	8.27	0.79	0.50
Recycling components	8	8.36	0.67	0.50
Lobbying	8	7.64	0.92	0.50
Taking back used products, components or materials	8	8.00	0.45	0.00
Providing used products, components or materials	8	8.18	0.60	0.50
Providing access to a product functionality	8	8.18	0.60	0.50

Note IQD = interquartile deviation; SD = standard deviation

(VHS: 9.00–10.00), while the remaining five were classified as highly significant (HS: 7.99–8.00). Notably, none of the variables were deemed insignificant within this framework. Furthermore, all variables achieved consensus, as indicated by their IQD values.

The eight sub-feature, cost structure, comprises 26 variables. As shown in Table 8.8, the panel of experts evaluating the CEBM framework for construction organisations rated fourteen of these variables as having very high significance (VHS: 9.00–10.00), while the remaining twelve were classified as having high significance (HS: 7.00–8.99). Notably, none of the variables were considered insignificant within the context of the CEBM framework for construction organisations. Also, consensus was reached for all variables except the cost of transportation and logistics, as indicated by their IQD values. However, the mean rating for the cost of transportation and logistics is 8.45, and the SD is 1.04.

The ninth sub-feature, revenue stream, comprises 20 variables. As shown in Table 8.9, the panel of experts evaluating the CEBM framework for construction organisations rated six of these variables as having very high significance (VHS: 9.00–10.00), while the remaining fourteen were classified as having high significance (HS: 7.00–8.99). Notably, none of the variables were considered insignificant within the context of the CEBM framework for construction organisations. Also, consensus was reached for all variables, as indicated by their IQD values.

The following sub-feature is a takeback system encompassing twelve variables. As detailed in Table 8.10, the panel of experts evaluating the CEBM framework for construction organisations rated nine variables as having very high significance (VHS: 9.00–10.00), while the remaining three were classified as highly significant (HS: 7.99–8.00). Notably, none of the variables were deemed insignificant within

Table 8.7 Key partnerships

Features	Median	Mean	SD	IQD
Suppliers	9	8.82	0.40	0.00
Manufacturers	9	8.82	0.40	0.00
Wholesalers	8	7.91	0.83	0.00
Retailers	8	7.91	0.94	1.00
Service providers	9	8.82	0.40	0.00
Public institutions	9	8.82	0.40	0.00
Collectors of products, components and material waste	9	8.73	0.65	1.00
Project managers	8	8.18	0.75	0.00
Architects	9	8.82	0.60	0.50
Engineers	9	8.82	0.60	0.50
Construction managers	8	8.09	0.30	0.00
Facilities managers	8	8.27	0.79	0.50
Waste disposal managers	9	8.91	0.70	0.50

Note IQD = interquartile deviation; SD = standard deviation

Table 8.8 Cost structure

Features	Median	Mean	SD	IQD
Cost of waste handling	9	8.82	0.60	0.50
Cost of waste processing	8	8.18	0.75	0.00
Cost of repair	8	8.27	0.65	1.00
Cost of maintenance	9	9.00	0.45	0.00
Cost of remanufacturing	9	8.91	0.54	0.00
Cost of resource inputs	9	8.82	0.40	0.00
Cost of transportation and logistics	9	8.45	1.04	1.50
Cost of supply risk	8	8.27	0.79	0.50
Cost of labour	9	8.73	0.79	0.50
Cost of facilitating collaboration	8	8.45	0.82	0.50
Cost of take-back management	8	8.00	1.00	0.00
Cost of client education	8	8.09	0.83	0.00
Cost of upgrading	9	8.82	0.60	0.50
Cost of designing out waste	9	8.91	0.54	0.00
Cost of refurbishing products/components	8	8.18	0.40	0.00
Cost of recycling products/components	9	8.91	0.54	0.00
Cost of energy recovery	9	9.00	0.45	0.00
Cost of producing long-lasting products/components	8	8.18	0.40	0.00
Cost of materials waste used for production input	8	8.18	0.40	0.00
Cost of chemical leasing	8	8.00	0.63	0.00
Cost of product/component renting	8	8.09	0.54	0.00
Cost of product/component leasing	8	8.09	0.70	0.00
Cost of producing on-demand	9	9.00	0.63	0.00
Cost of lean thinking	9	8.91	0.54	0.00
Cost of products/components upcycling	9	9.00	0.63	0.00
Cost of products/components downcycling	9	8.64	0.81	1.00

Note IQD = interquartile deviation; SD = standard deviation

this framework. Furthermore, all variables achieved consensus, as indicated by their IQD values.

The sub-feature consisted of thirteen variables for the material information management system attribute. Among these, six variables were identified as having very high significance (VHS: 9.00–10.00) in influencing CEBM for construction organisations. The remaining seven variables were classified as having high significance (HS: 7.00–8.99). None of the variables were considered insignificant, as shown in Table 8.11. Also, based on the IQD values, all the variables achieved a consensus level.

Table 8.9 Revenue stream

Features	Median	Mean	SD	IQD
Product service systems				
Pay per service/product/component	8	7.64	0.67	0.50
Availability based (subscription)	8	8.00	0.45	0.00
Usage-based (pay-per-use)	8	7.91	0.30	0.00
Performance-based (solution-oriented, demand fulfilment-oriented)	9	8.82	0.40	0.00
Outsourcing	8	8.09	0.30	0.00
Others (payment received for)				
Product/component lease	8	7.91	0.54	0.00
Product/component renting	8	7.91	0.54	0.00
Building maintenance	9	8.73	0.47	0.50
Repair of products/components	9	8.73	0.47	0.50
Upgrading	9	8.73	0.47	0.50
Remanufacturing components	8	8.36	0.67	0.50
Transformation	8	8.18	0.75	0.00
Waste handling processing	9	8.91	0.30	0.00
Client education	9	8.91	0.54	0.00
Designing out waste	8	8.27	0.79	0.50
Refurbishing products/components	8	8.27	0.65	0.00
Recycling products/components	8	8.09	0.30	0.00
Producing on demand	8	8.36	0.67	0.50
Products/components upcycling	8	8.18	0.40	0.00
Products/components downcycling	8	8.09	0.54	0.00

Note IQD = interquartile deviation; SD = standard deviation

The twelfth sub-feature, organisational culture, comprises 23 variables. As shown in Table 8.12, the panel of experts evaluating the CEBM framework for construction organisations rated twenty-two of these variables as having very high significance (VHS: 9.00–10.00), while the remaining one was classified as having high significance (HS: 7.00–8.99). Notably, none of the variables were considered insignificant within the context of the CEBM framework for construction organisations. Also, consensus was reached for all variables, as indicated by their IQD values.

Table 8.10 Takeback system

Features	Median	Mean	SD	IQD
Incentivised return	9	8.91	0.54	0.00
Collection of used products	8	8.18	0.60	0.50
Volumes of products/materials returned	9	8.73	0.47	0.50
Ease of takeback	9	8.73	0.47	0.50
Transportation system	9	8.82	0.60	0.50
Distance of takeback	8	7.82	0.75	0.00
Condition of returned material/product	8	7.82	0.75	0.00
Documentation	9	8.82	0.87	0.50
Recovery rate	9	8.91	0.83	0.00
Client satisfaction with the system	9	8.73	1.19	0.00
Environmental impact of system	9	9.00	0.77	0.00
Cost-effectiveness of the system	10	9.91	0.30	0.00

Note IQD = interquartile deviation; SD = standard deviation

Table 8.11 Material information management system

Features	Median	Mean	SD	IQD
Standardisation of all materials-related data	8	8.00	0.45	0.00
Centralisation of all materials-related data	8	8.00	0.63	0.00
Monitoring and managing inventory levels in real-time	9	8.73	1.01	0.00
Materials requirements planning	9	8.73	0.79	0.50
Creation of bills of materials	8	8.09	0.70	0.50
Management of bills of materials	8	8.18	0.60	0.50
Suppliers' information management	8	8.09	0.54	0.00
Suppliers' relationship management	8	8.00	0.63	0.00
Suppliers' performance management	8	8.00	0.63	0.00
Tracking suppliers' compliance with quality and regulatory requirements	9	8.82	0.60	0.50
Reporting and analytics	9	8.82	0.98	0.00
Integration with other systems	9	8.64	1.03	0.50
Material traceability	9	8.64	1.29	0.00

Note IQD = interquartile deviation; SD = standard deviation

Table 8.12 Organisational culture

Features	Median	Mean	SD	IQD
Clan culture				
Stakeholders' involvement in CEBM	9	9.00	0.45	0.00
Teamwork	9	9.00	0.45	0.00
Cooperation	9	9.00	0.45	0.00
Mentoring	9	9.00	0.45	0.00
Skills development acquisition	9	9.00	0.45	0.00
Adhocracy culture				
Reward incentive system	9	8.82	0.40	0.00
Entrepreneurship development	9	8.82	0.40	0.00
Motivation	9	8.82	0.40	0.00
Updated technology systems	9	9.09	0.54	0.00
Acquisition of new technology	9	8.82	0.40	0.00
Innovation	9	9.09	0.30	0.00
Hierarchy culture				
Leadership control	9	9.09	0.54	0.00
Regulation of activities	9	9.00	0.63	0.00
Assertiveness on policies	9	9.00	0.63	0.00
Efficiency among project team members	9	9.00	0.45	0.00
Meeting targets with adherence to schedules	9	9.00	0.45	0.00
Smooth CEBM activity operations	9	8.91	0.54	0.00
Prediction of activity with the CEBM	9	9.00	0.45	0.00
Market culture				
Client satisfaction	9	9.09	0.54	0.00
Result-oriented	9	9.09	0.54	0.00
Emphasis on goal accomplishment	9	8.91	0.30	0.00
Healthy competition	8	8.27	0.47	0.50
Ensuring a large market size	9	8.82	0.40	0.00

Note IQD = interquartile deviation; SD = standard deviation

8.2.3 DSO3: To Predict the Critical Factors that Could Affect the Adoption of a CEBM Framework in Construction Organisations

Several critical factors can influence the adoption of a CEBM framework in construction organisations. These factors can be pivotal in determining how effectively the framework can be integrated into existing operations and sustained over time. One of the foremost considerations is the economic viability of circular practices. For many construction organisations, transitioning to a circular model requires significant upfront investment in new technologies, materials, and processes. However, the return on investment is often long-term, forcing firms to balance short-term financial pressures with the promise of future economic benefits. Based on these, it is essential to explore the extent to which critical factors may influence the adoption of a CEBM framework within construction organisations.

The study identified 24 critical factors that may influence the adoption of the CEBM framework in construction organisations. These factors were then presented to a panel of experts to evaluate their significance and determine the extent of their impact on the adoption of the CEBM framework. As shown in Table 8.13, the panel of experts rated eleven of these variables as having very high significance (VHS: 9.00–10.00), while the remaining thirteen were classified as having high significance (HS: 7.00–8.99). Notably, none of the variables were considered insignificant in the critical factors that may influence the adoption of the CEBM framework in construction organisations. Also, consensus was reached for all variables, as indicated by their IQD values.

8.2.4 DSO4: To Determine the Extent to Which Policies Can Drive the Adoption of a CEBM Framework in Construction Organisations

The construction industry plays a crucial role in advancing the principles of CE. Transitioning to CEBM requires strong policy frameworks that create enabling conditions for change. Several policies can drive this transformation by addressing the sector's economic, regulatory, and behavioural challenges. For example, governments can establish strict regulations to encourage construction organisations to adopt circular practices. Furthermore, adopting a CEBM framework in construction organisations is contingent upon a cohesive blend of policies. Policymakers can encourage construction firms to rethink their approaches, design for circularity, and ultimately contribute to a more sustainable built environment by creating a supportive ecosystem. Such policies can benefit the environment and enhance industry resilience and competitiveness in a resource-constrained world.

The study identified 13 policies that may influence the adoption of the CEBM framework in construction organisations. These policies were then presented to a panel of experts to evaluate their significance and determine the extent of their impact on adopting the CEBM framework. As shown in Table 8.14, the panel of experts rated seven of these variables as having very high significance (VHS: 9.00–10.00), while the remaining six were classified as having high significance (HS: 7.00–8.99). Notably, none of the policies were considered insignificant in the policies that may influence the adoption of the CEBM framework in construction organisations. Also, consensus was reached for all variables, as indicated by their IQD values.

Table 8.13 Critical factors

Factors	Median	Mean	SD	IQD
Ineffective C&D waste dismantling processes	8	8.00	0.45	0.00
Ineffective C&D sorting processes	8	8.00	0.45	0.00
Ineffective C&D recovering processes	8	8.00	0.45	0.00
Not green designing of construction projects	8	8.09	0.54	0.00
Using finitely recyclable construction materials	8	8.18	0.40	0.00
Overemphasising non-environmental friendly methods	8	8.18	0.60	0.50
Preferring off-site C&D waste sorting/landfilling over on-site sorting	8	7.91	0.54	0.00
Inadequate policies to manage C&D wastes	9	8.73	0.65	1.00
Inadequate legal frameworks to manage C&D wastes	9	8.82	0.60	0.50
Lack of supervision on C&D waste management	8	8.09	0.70	0.00
Lack of producer-based responsibility system in the production of construction materials	8	8.00	0.45	0.00
Lack of clearly defined national goals to adopt circular economy	9	8.55	0.93	0.50
Lack of awareness of circular economy	9	8.64	1.03	0.50
The inherent complexity of transforming to a circular economy in C&D waste management	8	8.36	0.92	0.50
Lack of integration of sustainable C&D waste management	9	8.82	0.40	0.00
Risk aversion	9	8.82	0.40	0.00
User preference for new construction materials over reused/ recycled ones	8	8.00	0.63	0.00
Uncertain aftermaths of moving toward circular economy	9	8.82	0.40	0.00
Non-standardized C&D waste reduction reporting	9	8.73	0.65	0.00

(continued)

Table 8.13 (continued)

Factors	Median	Mean	SD	IQD
Lack of funding to implement circular economy	9	8.91	0.70	0.00
Tendency to manage cost and time rather than C&D wastes	9	8.55	0.69	1.00
Agency and ownership issues in C&D waste management	9	8.73	0.65	0.00
Lack of commitment by top urban managers	8	8.09	0.54	0.00
Ineffective C&D waste management	8	8.00	0.77	0.00

Note IQD = interquartile deviation; SD = standard deviation; C&D = construction and demolition

Table 8.14 Policies

Policies	Median	Mean	SD	IQD
Financial subsidies for recycling	9	8.82	0.60	0.50
Tax reliefs	8	8.09	0.54	0.00
Incentives for consumers to purchase recycled materials	8	8.00	0.45	0.00
Waste management regulations	9	9.09	0.30	0.00
Fixed non-recycling tax	9	8.64	0.50	1.00
Support for the construction of CDW processing facilities	8	8.18	0.40	0.00
Landfill charging schemes	9	8.55	0.52	1.00
Taxes on raw materials	8	7.82	0.75	0.00
Construction waste site management plans	9	8.64	0.67	0.50
Products standards	9	8.73	0.79	0.50
Waste facility permit scheme	8	8.00	0.45	0.00
Incentives for innovative green technology	9	8.91	0.54	0.00
Minimum recycling rate target	8	8.18	0.60	0.50

Note IQD = interquartile deviation; SD = standard deviation

8.2.5 DSO5: To Identify the Extent to Which the Benefits of a CEBM Framework Can Drive the Adoption of a CEBM Framework in Construction Organisations

Adopting the CEBM framework in construction organisations offers numerous benefits that align with environmental and economic objectives, driving industry-wide transformation. By integrating CE principles, construction organisations can contribute to reducing landfill use and mitigating environmental degradation. The transition to a CEBM framework can offer construction organisations opportunities to reduce costs, innovate, improve stakeholder relations, and align with emerging regulations while making a tangible contribution to global sustainability goals.

The study identified 23 benefits of adopting the CEBM framework in construction organisations. These benefits were then presented to a panel of experts to evaluate their significance and determine the extent of their impact on adopting the CEBM framework. As shown in Table 8.15, the panel of experts rated nineteen of these benefits as having very high significance (VHS: 9.00–10.00), while the remaining four were classified as having high significance (HS: 7.00–8.99). Notably, none of the policies regarding the benefits of adopting the CEBM framework in construction organisations were considered insignificant. Also, consensus was reached for all variables, as indicated by their IQD values.

Table 8.15 CEBM benefits

Benefits	Median	Mean	SD	IQD
Environmental gain	9	9.00	0.45	0.00
Economic gain	9	8.82	0.75	0.00
Resource optimisation	9	8.91	0.54	0.00
Collaboration among stakeholders	9	8.82	0.40	0.00
Extended producer responsibility	9	8.73	0.47	0.50
Eco-friendly labelling	9	9.09	0.54	0.00
Promote industrial upgrading	9	8.82	0.40	0.00
Improve energy efficiency	9	9.09	0.54	0.00
Sustainable innovation	9	8.91	0.54	0.00
Connectivity of material streams	8	8.27	0.79	0.50
Quick response to market demand	9	8.91	0.70	0.00
Raw materials reduction	8	8.27	0.90	1.00
Waste reduction	9	9.09	0.54	0.00
Air emissions reduction	9	8.91	0.83	0.00
Water effluents reduction	9	8.82	0.60	0.50
Profitability	8	8.09	0.54	0.00
Variable cost reduction	9	8.73	0.47	0.50
Increased revenues	9	8.64	0.67	0.50
Increased market share	9	8.64	0.50	1.00
Economic efficiency	9	8.91	0.30	0.00
Improved productivity	9	8.91	0.30	0.00
Increased employment	9	8.82	0.60	0.50
Community development	9	9.00	0.63	0.00

Note IQD = interquartile deviation; SD = standard deviation

8.3 Discussion of Results

8.3.1 DSO1 to Determine the Main Attributes of CEBM for Construction Organisations

The first objective of the Delphi study was to uncover the essential components of a CEBM framework tailored for construction organisations. The findings from the Delphi process revealed that all twelve identified features achieved consensus, underscoring their importance. This outcome aligns with previous research on CEBM frameworks [6, 20, 24]. The features received median scores ranging from high significance to very high significance, and their IQD values satisfied the threshold for consensus. These results highlight that value proposition, client segment, channels, client relationship, key resources, key activities, key partnerships, cost structure, revenue streams, takeback system, material information management systems, and organisational culture are critical elements in developing a CEBM framework for construction organisations.

8.3.2 DSO2 to Determine the Sub-attributes of CEBM for Construction Organisations

The second objective of the Delphi study was to determine the sub-attributes that contribute to the framework of CEBM for construction organisations. The results from the Delphi process indicate that most sub-attributes related to the framework of CEBM in construction are highly significant. Regarding the value proposition, all 22 identified sub-attributes were considered significant. Among these, seven were rated as having very high significance, while the remaining 15 were deemed highly significant. Furthermore, consensus was reached for all 22 attributes except one (waste handling and processing), as their IQD values fell within the predefined cut-off range. These findings align with several studies that emphasise the significance of many value proposition sub-attributes as key components of the CEBM framework [17, 18, 27, 34]. For instance, Zhang et al. [41] suggest that downcycling in the construction sector is feasible, mainly through backfilling concrete waste generated from construction and demolition activities. Monier et al. [29] further elaborate that downcycling can be carried out either off-site at a stationary facility or on-site using mobile crushers.

Experts also view value propositions such as producing durable construction components, upgrading buildings, managing construction waste, and processing waste as viable approaches for the construction industry. Durable construction components can enhance structures' longevity, safety, and sustainability. However, achieving this involves implementing rigorous quality control processes during production, including regular inspections, testing, and strict adherence to industry standards. This approach ensures that defects or deviations are promptly addressed while emphasising designs that facilitate easy maintenance. By focusing on innovative design, environmental sustainability, and compliance with standards, manufacturing durable components can lead to structures capable of enduring over time. Moreover, effectively handling and processing construction waste is vital for sustainable construction practices. Construction and demolition waste pose significant global challenges to the industry's performance [28]. Proper waste management practices, such as segregation, collection, and transportation, can reduce environmental harm, improve resource efficiency, and ensure adherence to waste disposal regulations. These measures contribute to minimising the adverse impacts of waste on both the environment and industry performance.

Furthermore, within the client segment attribute, which consists of seven sub-attributes, three were identified as being of very high significance, while the remaining four were considered highly significant. Moreover, all sub-attributes demonstrated a strong level of consensus, with IQD values between 0.00 and 0.50. These findings align with previous research, such as Lewandowski [20], which also emphasises the relevance of the client segment in the CEBM framework. According to Lewandowski [20], the client segment is closely associated with the value proposition component. Value proposition fit is illustrated by the value proposition design for the client segment. Quality and cost-conscious clients will prioritise the quality of a circular product or service and the associated costs. These clients will seek a balanced approach that ensures they receive value for their investment. Quality and cost-conscious clients focus on the concept of value. They aim to maximise the value they receive by obtaining high-quality products or services at a reasonable cost. This involves finding the right balance between quality and expenditure. Quality and cost-conscious construction clients will navigate the complex landscape of procurement and decision-making by carefully balancing their desire for high-quality construction with a pragmatic approach to project costs. The positive aspect of this mindset is that it contributes to sustainable and successful business practices, driving enduring relationships with suppliers, providers, and end-users.

For channels attribute, the category included 13 sub-attributes, of which only one (web advertisements) was deemed to have very high significance. At the same time, 11 were rated as highly significant, and one was rated with medium significance. Despite this variation, all sub-attributes reached consensus, as indicated by their IQD values exceeding the established threshold. Among these, retailers, websites, web advertisements, direct personal selling, and newsletters stood out by achieving a very high consensus, with IQD values at 0.00. Expert evaluations of channel attributes indicated that value propositions aimed at construction clients can be effectively delivered through direct personal selling, social media, and word-of-mouth. Considering the nature of construction industry products, these items cannot be provided virtually [20]. Therefore, construction organisations can utilise the three identified channels primarily for communication. This approach aligns with insights from EMF [5], highlighting that virtualisation significantly transforms channels to support circular business models.

The evaluation of the sub-attributes related to client relationships identified 13 distinct sub-features. Among these, five were found to be of very high significance, while the remaining eight were deemed highly significant. Furthermore, there was consensus among the Delphi experts, as all sub-features IQD values fell within the acceptable threshold. These findings align with previous research, such as Lewandowski [20], which also emphasises the relevance of the client relationship in the CEBM framework. Lewandowski [20] suggests that building and maintaining strong client relationships can be pivotal to the core concept of CE. The experts-identified sub-attributes mentioned could support construction organisations in achieving this objective.

Furthermore, within the key resources attribute, which consists of seven sub-attributes, four were identified as being of very high significance, while the remaining three were considered highly significant. Moreover, all sub-attributes demonstrated a strong level of consensus, with IQD values between 0.00 and 0.50. These findings align with previous research [11]. When evaluating the key resource sub-attributes necessary for implementing a CEBM in construction organisations, the Delphi experts emphasise the significance of circular sourcing and marketplace practices. Another crucial sub-attribute identified is the replacement of traditional resources with materials that exhibit superior performance during construction. This substitution is significant because conventional construction materials require substantial thermal and electrical energy for manufacturing [11]. According to Maraveas [26], the production processes for these traditional materials contribute significantly to environmental pollution, affecting air, land, and water and leading to a higher carbon footprint. Replacing traditional materials with more efficient and environmentally friendly alternatives can enhance construction projects' quality, sustainability, and overall efficiency. These alternative materials are designed to offer improved characteristics, such as excellent durability, strength, energy efficiency, and reduced environmental impact. However, adopting such materials requires careful consideration of various factors, including technical feasibility, environmental implications, economic viability, and regulatory compliance. This transition also necessitates stakeholder collaboration and a strong commitment to sustainable, high-quality construction practices.

The evaluation of the sub-attributes related to key activities identified 13 distinct sub-features. Among these, four were found to be of very high significance, while the remaining nine were deemed highly significant. Furthermore, there was consensus among the Delphi experts, as all sub-features IQD values fell within the acceptable threshold. These findings align with previous research, such as Lewandowski [20], which also emphasises the relevance of the key activities in the CEBM framework. The application of CE principles can be integrated into various activities that either directly or indirectly contribute to creating, delivering, and supporting value propositions [20]. The process control sub-attribute within construction organisations encompasses the planning, managing, and optimising of construction delivery processes to achieve optimal efficiency and performance. In addition, extending the lifespan of products and components involves adopting methods and strategies designed to enhance the durability, functionality, and longevity of construction materials, products, and components. This approach is vital for promoting sustainable construction, improving cost-efficiency, and reducing environmental impact. By adopting these practices, construction organisations can support sustainable development, minimise their ecological footprint, and maximise the lifecycle performance of their projects.

For key partnerships attribute, the category included 13 sub-attributes, of which eight were deemed to have very high significance, while the remaining five were rated as highly significant. Despite this variation, all sub-attributes reached consensus, as indicated by their IQD values exceeding the established threshold. Among these, suppliers, manufacturers, wholesalers, service providers, public institutions, project managers, and construction managers stood out by achieving very high consensus, with IQD values recorded at 0.00. It is evident from the findings that construction organisations can gain significant advantages by sourcing materials through collaborative networks, which play a crucial role in improving process operations and management. Collaboration helps secure vital resources and execute key tasks more efficiently. For instance, stakeholders can manage offsite recycling, handling construction and demolition waste once it is no longer needed, or even recycling materials sold to other industries. However, the absence of collaboration, or insufficient collaboration, is one of the most frequent professional challenges that impact the execution of construction projects, alongside other limitations faced by industry experts [4]. Therefore, construction stakeholders must embrace a collaborative approach to implement the CEBM within the construction sector effectively.

Furthermore, the cost structure attribute consists of twenty-six sub-attributes. Fourteen of the sub-attributes were identified as being of very high significance, while the remaining twelve were considered highly significant. Moreover, all sub-attributes demonstrated a strong level of consensus except for the cost of transportation and logistics. These findings support Lewandowski [20] argument that organisations may need to implement further adjustments whenever they decide to change their cost structure. Considering that decisions made during the planning and design stages contribute to the amount of waste generated in the construction industry [3, 31], construction organisations must undergo substantial changes to eliminate waste in construction projects. The costs associated with waste reduction in design are complex, encompassing factors like material selection, process optimisation, regulatory adherence, brand impact, and long-term sustainability. Although there may be upfront costs, the financial rewards and environmental and social benefits make waste elimination in design a strategic business approach in today's industry.

For revenue streams attribute, the category included 20 sub-attributes, of which six were deemed to have very high significance, while the remaining 14 were rated as highly significant. Despite this variation, all sub-attributes reached consensus, as indicated by their IQD values exceeding the established threshold. The Delphi experts clearly identified pay-per-service/component and performance-based payment as highly significant when evaluating the revenue stream sub-attributes within the CEBM framework for construction organisations. Essentially, revenue streams represent the income sources for construction firms. Several CEBMs, particularly those related to product-service systems [39], are applicable here. An example of such a model is the pay-per-service/component approach. The Delphi experts suggest that construction organisations could effectively adopt this revenue model. Recognising that value propositions are key in shaping revenue sources is also crucial.

Furthermore, the Delphi experts assessed take-back sub-attributes. This attribute has 12 sub-attributes, of which nine were deemed to have very high significance, while the remaining three were rated as highly significant. Despite this variation, all sub-attributes reached consensus, as indicated by their IQD values exceeding the established threshold. Thus, it can be concluded that for construction organisations to implement such a system successfully, cost-effectiveness and client satisfaction must be prioritised. According to EMF [5], the core concept of the CE revolves around material loops, making reverse logistics (take-back systems) essential for the realisation of a CEBM in the construction sector.

Evaluating the sub-attributes related to the material information management system identified 13 distinct sub-features. Among these, six were found to be of very high significance, while the remaining seven were deemed highly significant. Furthermore, there was consensus among the Delphi experts, as all sub-features IQD values fell within the acceptable threshold. In Lewandowski's [20] conceptual framework for CEBMs, adopting suitable information technology (IT) and data management systems was highlighted as essential for enabling material tracking. Similarly, Scott [36], Laubscher and Marinelli [19], and Nilsson and Söderberg [32] emphasised the need for Material Information Management Systems (MIMS). The study argued that MIMS, alongside the traditional components of the business model canvas, is a critical factor in developing CEBMs. Unlike the traditional canvas, which primarily focuses on profit generation, MIMS supports the broader objectives of CEBMs. Peters et al. [33] further argued that incorporating MIMS is indispensable in CEBMs for the construction industry, as it facilitates the modeling, tracking, capturing, and management of material data throughout the material lifecycle. The Delphi experts on this attribute underscored the significance of MIMS in CEBMs for construction organisations. These systems enable firms to align with CE principles while enhancing their financial performance. By leveraging MIMS, construction organisations can gain a competitive advantage, leading to better financial outcomes. MIMS also supports these organisations in building stronger relationships with stakeholders, increasing motivation, satisfaction, and loyalty among partners and clients. The findings suggest that MIMS is pivotal in driving sustainability within the construction sector. It helps reduce costs, conserve resources, and lower energy consumption. In addition, it contributes to improving organisational and managerial capabilities while encouraging the adoption of renewable energy sources and environmentally friendly practices. Lastly, MIMS benefits businesses and the environment by driving innovation and operational efficiency.

Furthermore, evaluating the sub-attributes related to organisational culture identified 23 distinct sub-features. Among these, 22 were found to be of very high significance, while only one was deemed highly significant. Furthermore, there was a very strong consensus among the Delphi experts, as 22 sub-features have IQD values of 0.00. Only one sub-attribute had a 0.50 IQD value, which signifies a strong consensus.

This implies that organisational culture is a unique aspect of every organisation, characterised by its distinct set of beliefs, values, rituals, symbols, artefacts, and guidelines that define acceptable employee behaviour. Globalisation and digitisation are two significant forces driving organisations to adapt and reform their cultures. However, such changes often encounter overt or covert resistance from employees. This potentially leads to failure if they conflict with deeply rooted cultural norms [30]. Shared values and beliefs are central to organisational culture, and understanding them is often seen as essential for being part of the organisation [7]. Delphi experts' insights underscore organisational culture's critical role in adopting CEBMs in construction organisations. Soft management factors like culture are indispensable for these organisations to transition successfully to CE practices. Human-centric elements, such as shared values and norms, significantly influence the adoption of CEBMs at the corporate level [2]. To enable CE principles to transform construction processes fundamentally, organisations must revise their existing norms, visions, tools, and performance indicators [15].

8.3.3 DSO3: To Predict the Critical Factors that Could Affect the Adoption of a CEBM Framework in Construction Organisations

The third objective of the Delphi study was to predict the critical factors that could affect the adoption of a CEBM framework in construction organisations. The results from the Delphi process indicate that all critical factors that could affect the adoption of a CEBM framework in construction organisations are highly significant. All 24 identified critical factors were considered significant. Among these, eleven were rated as having very high significance, while the remaining 13 were deemed highly significant. Furthermore, consensus was reached for all 24 factors, as their IQD values fell within the predefined cut-off range.

The Delphi experts' session identified insufficient CE-related policies as a critical factor affecting the adoption of CEBM in construction organisations. Other critical factors include a lack of a robust legal framework and inadequate funding to implement CE principles. Kabirifar et al. [14] suggest that developed and developing economies can leverage various policy tools to promote recycling. Effective CE policies are essential for advancing sustainable resource use, reducing waste, and conserving the environment. For example, these policies can prioritise reusing and recycling materials from demolished structures, turning waste into valuable resources for future construction projects. This approach can also reduce reliance on virgin materials, among other benefits. Numerous studies [8, 13, 25, 38] have emphasised the critical role of governments and regional authorities in facilitating the transition to CEBM. Governments can play a pivotal role by providing essential technology and infrastructure [10, 25], introducing new legislation and regulations, and implementing eco-labels [35]. They can further advance circularity through public procurement policies and support for circular business initiatives [40]. In addition, governments are vital in coordinating multi-stakeholder processes and enforcing regulations across supply chain participants and competitors.

Furthermore, Tsoi [37] advocates collaboration among local governments, NGOs, and multinational corporations (MNCs) to enhance CE adoption. Local governments, such as municipalities, are essential in planning, coordinating, and managing transportation, recycling, and waste disposal [16]. In conclusion, the findings highlight the critical factors that could affect the adoption of CEBM in construction organisations. Addressing these factors could also resolve broader challenges, paving the way for a more sustainable industry.

8.3.4 DSO4: To Determine the Extent to Which Policies Can Drive the Adoption of a CEBM Framework in Construction Organisations

The fourth objective of the Delphi study was to determine the extent to which policies can drive the adoption of a CEBM framework in construction organisations. All 13 identified policies were considered significant. Among these, seven were rated as having very high significance, while the remaining six were deemed highly significant. Furthermore, consensus was reached for all 13 policies, as their IQD values fell within the predefined cut-off range. The Delphi experts' findings show that financial subsidies for recycling are a key policy to drive the adoption of the CEBM framework in construction organisations. In addition, waste management regulations and tax incentives were identified as important supporting measures. Providing financial support is a common and effective strategy to promote the growth of the recycling sector. For example, some countries offer tax exemptions, low-interest loans to businesses engaged in recycling construction materials, and tax benefits to consumers purchasing recycled products [9]. In Singapore, the government established a recycling-focused industrial park, renting out facilities to recycling businesses at reduced rates to encourage participation in sustainable practices [21, 23]. These initiatives aim to motivate construction organisations and contractors to adopt and invest in recycling technologies, thereby mitigating the environmental impact of construction activities.

The Delphi study also suggests that grants for establishing and enhancing recycling facilities could significantly support the industry. Construction firms planning to invest in advanced equipment for sorting, crushing, and processing construction and demolition waste could apply for grants to offset capital expenses. This financial assistance will ease the cost burden on businesses and drive the development of an efficient recycling infrastructure. Direct subsidies may also be provided for activities like transporting recycled materials or purchasing recycled products, reducing the cost disparity between traditional waste disposal and recycling solutions. This makes recycling a more financially attractive option for businesses. Tax incentives emerged as another effective tool to encourage recycling in the construction sector. As recognised by the Delphi experts, governments could offer tax credits or deductions to construction companies that integrate recycling into their waste management practices. These incentives will provide tangible benefits for companies demonstrating environmental responsibility, driving a culture of sustainability within the industry. In summary, financial support mechanisms such as subsidies, grants, and tax incentives can significantly enhance the adoption of the CEBM framework in construction organisations. These measures cascade effect throughout the supply chain, encouraging material manufacturers to incorporate recycled content into their products. Ultimately, such policies will create a more sustainable construction sector, reducing the ecological footprint associated with construction and demolition activities.

8.3.5 DSO5: To Identify the Extent to Which the Benefits of a CEBM Framework Can Drive the Adoption of a CEBM Framework in Construction Organisations

The fifth objective of the Delphi study was to identify the extent to which the benefits of a CEBM framework can drive the adoption of a CEBM framework in construction organisations. All 23 identified benefits were considered significant. Among these, twenty were rated as having very high significance, while the remaining three were deemed highly significant. Furthermore, consensus was reached for all 23 policies, as their IQD values fell within the predefined cut-off range. The findings indicate that embracing the CEBM in construction organisations offers numerous benefits. One notable benefit is the potential to reduce greenhouse gas emissions. The construction industry accounts for approximately 40% of global energy consumption and 33% of greenhouse gas emissions [12, 22], and implementing the CEBM framework could significantly lower these figures. This perspective aligns with the insights shared by Delphi experts.

In addition, the experts highlighted that adopting the CEBM framework could enhance profitability for construction organisations. This increase in revenue would stem from the emergence of new business opportunities, job creation, the development of innovative value propositions, and the expansion of service-oriented revenue streams. Through practices such as reusing and recycling construction materials, components, and items, organisations can reduce reliance on virgin resources and achieve cost savings. These findings align with Agyemang et al. [1], who identified cost reduction as a key driver of CE adoption. Hence, construction organisations may be motivated to transition to CEBM by minimising material costs and mitigating risks associated with resource scarcity and price fluctuations, particularly given the sector's high resource demands. In summary, shifting from a linear "take-make-dispose" production and consumption model to the CEBM enables construction organisations to reduce waste and pollution, preserve natural resources, and unlock new economic opportunities.

8.4 Summary

This chapter presented the findings from the first and second rounds of the Delphi study. For each question posed to the expert panel, detailed computations were conducted. These questions addressed the importance of the main and sub-attributes within the circular economy business model (CEBM) framework for construction organisations. In addition, the study examined the critical factors that could affect the adoption of a CEBM framework in construction organisations. It also explored the degree to which policies can drive the adoption of a CEBM framework in construction organisations. The chapter concluded with an in-depth discussion of the results of the study's objectives. The findings highlighted a clear consensus among the experts on the main attributes of the CEBM framework for construction organisations. Ultimately, the results identified critical constructs essential for developing the CEBM framework for construction organisations.

References

1. Agyemang M, Kusi-Sarpong S, Khan SA, Mani V, Rehman ST, Kusi-Sarpong H (2019) Drivers and barriers to circular economy implementation: an explorative study in Pakistan's automobile industry. Manag Decis 57(4):971–994
2. Chiappetta Jabbour CJ, Sarkis J, de Sousa L, Jabbour AB, Scott Renwick DW, Singh SK, Grebinevych O, Kruglianskas I, Filho MG (2019) Who is in charge? A review and a research agenda on the 'human side' of the circular economy. J Clean Prod 222:793–801
3. Crawford RH, Mathur D, Gerritsen R (2017) Barriers to improving the environmental performance of construction waste management in remote communities. Procedia Eng 196:830–837
4. Deep S, Gajendran T, Jefferies M (2021) A systematic review of 'enablers of collaboration'among the participants in construction projects. Int J Constr Manag 21(9):919–931
5. Ellen MacArthur Foundation (2015) Growth within: a circular economy vision for a competitive Europe, vol 100. Ellen MacArthur Foundation, pp 81–92
6. Ellen MacArthur Foundation, & Ide (2017) Business model canva. http://circulardesignguide.com/post/circular-business-model-canvas
7. Emmanuel O (2017) Understanding organisational culture and organisational performance: are they two sides of the same coin? J Manage Res 9(1):12–21
8. Galvão GDA, De Nadae J, Clemente DH, Chinen G, De Carvalho MM (2018) Circular economy: overview of barriers. Procedia Cirp 73:79–85
9. Gao MZA (2018) Construction and demolition waste management: from Japan to Hong Kong. Griffin's View on International and Comparative Law, pp 1–29
10. Geng Y, Doberstein B (2008) Developing the circular economy in China: challenges and opportunities for achieving 'leapfrog development.' Int J Sustain Develop World Ecol 15(3):231–239
11. Guna V, Ilangovan M, Rather MH, Giridharan BV, Prajwal B, Vamshi Krishna K, Venkatesh K, Reddy N (2020) Groundnut shell/rice husk agro-waste reinforced polypropylene hybrid biocomposites. J Build Eng 27:100991
12. Hong J, Shen Q, Xue F (2016) A multi-regional structural path analysis of the energy supply chain in China's construction industry. Energy Policy 92:56–68
13. Jia F, Zuluaga-Cardona L, Bailey A, Rueda X (2018) Sustainable supply chain management in developing countries: an analysis of the literature. J Clean Prod 189:263–278

14. Kabirifar K, Mojtahedi M, Wang CC, WY TV (2020) A conceptual foundation for effective construction and demolition waste management. Cleaner Eng Technol 1:100019
15. Korhonen J, Honkasalo A, Seppälä J (2018) Circular economy: the concept and its limitations. Ecol Econ 143:37–46
16. Kunz N, Mayers K, Van Wassenhove LN (2018) Stakeholder views on extended producer responsibility and the circular economy. Calif Manage Rev 60(3):45–70
17. Lacy P, Keeble J, McNamara R, Rutqvist J, Haglund T, Cui M, Cooper A, Pettersson C, Kevin E, Buddemeier P (2014) Circular advantage: innovative business models and technologies to create value in a world without limits to growth. Accenture
18. Lacy P, Long J, Spindler W, Lacy P, Long J, Spindler W (2020) The circular business models. In: The circular economy handbook: realizing the circular advantage. pp 17–42
19. Laubscher M, Marinelli T (2014) Integration of circular economy in business. In: Proceedings of the conference: going Green—Care Innovation 2014, Vienna, Austria, 17–20 November 2014
20. Lewandowski M (2016) Designing the business models for circular economy—towards the conceptual framework. Sustainability 8(1):43
21. Li J, He G, Zhong X (2017) Research on policy instruments selection of construction and demolition waste recycling management in Japan, Germany and Singapore. Constr Econ 38(5):87–90
22. Li YL, Han MY, Liu SY, Chen GQ (2019) Energy consumption and greenhouse gas emissions by buildings: a multi-scale perspective. Build Environ 151:240–250
23. Li J, Yao Y, Zuo J, Li J (2020) Key policies to the development of construction and demolition waste recycling industry in China. Waste Manage 108:137–143
24. Lüdeke-Freund F, Gold S, Bocken NM (2019) A review and typology of circular economy business model patterns. J Ind Ecol 23(1):36–61
25. Manninen K, Koskela S, Antikainen R, Bocken N, Dahlbo H, Aminoff A (2018) Do circular economy business models capture intended environmental value propositions? J Clean Prod 171:413–422
26. Maraveas C (2020) Production of sustainable construction materials using agro-wastes. Materials 13(2):262
27. Mentink B (2014) Circular business model innovation: a process framework and a tool for business model innovation in a circular economy. http://resolver.tudelft.nl/uuid:c2554c91-8aaf-4fdd-91b7-4ca08e8ea621
28. Mohammed M, Shafiq N, Abdallah NAW, Ayoub M, Haruna A (2020, April) A review on achieving sustainable construction waste management through application of 3R (reduction, reuse, recycling): a lifecycle approach. In: IOP conference series: earth and environmental science, vol 476, no 1. IOP Publishing, p 012010
29. Monier V, Hesstin M, Impériale A, Prat L, Hobbs G, Ramos KAM (2017) Resource efficient use of mixed wastes: improving management of construction and demolition waste. European Union. https://op.europa.eu/en/publication-detail/-/publication/78e42e6cd8a6-11e7-a506-01aa75ed71a1/language-en
30. Mory L, Wirtz BW, Göttel V (2016) Factors of internal corporate social responsibility and the effect on organizational commitment. Int J Hum Resour Manag 27(13):1393–1425. https://doi.org/10.1080/09585192.2015.1072103
31. Newaz MT, Davis P, Sher W, Simon L (2022) Factors affecting construction waste management streams in Australia. Int J Constr Manag 22(13):2625–2633
32. Nilsson N, Söderberg V (2015) How to future proof a business model: capture and capitalize value in the field of urban mining. Master's thesis, Blekinge Institute of Technology, Karlskrona, Sweden
33. Peters M, Ribeiro A, Oseyran J, Wang K (2017) Buildings as material banks and the need for innovative business models. https://www.bamb2020.eu/topics/businessmodels/
34. Planing P (2015) Business model innovation in a circular economy reasons for non-acceptance of circular business models. Open J Bus Model Innovation 1(11):1–11

35. Prieto-Sandoval V, Jaca C, Ormazabal M (2018) Towards a consensus on the circular economy. J Clean Prod 179:605–615
36. Scott JT (2017) The sustainable business: a practitioner's guide to achieving long-term profitability and competitiveness. Routledge
37. Tsoi J (2010) Stakeholders' perceptions and future scenarios to improve corporate social responsibility in Hong Kong and Mainland China. J Bus Ethics 91:391–404
38. Van Buren N, Demmers M, Van der Heijden R, Witlox F (2016) Towards a circular economy: the role of Dutch logistics industries and governments. Sustainability 8(7):647
39. Van Renswoude K, Wolde AT, Joustra DJ (2015) Circular business models. Part 1: an introduction to IMSA's circular business model scan. https://groenomstilling.erhvervsstyrelsen.dk/sites/default/files/media/imsa_circular_business_models_-_april_2015_-_part_1.pdf
40. Yuan Z, Bi J, Moriguichi Y (2006) The circular economy: a new development strategy in China. J Ind Ecol 10
41. Zhang C, Hu M, Di Maio F, Sprecher B, Yang X, Tukker A (2022) An overview of the waste hierarchy framework for analyzing the circularity in construction and demolition waste management in Europe. Sci Total Environ 803:149892

Chapter 9
Conclusion

Abstract This book establishes a framework for a circular economy business model (CEBM) tailored to construction organisations, addressing the sector's unique challenges and opportunities. The framework identifies the main attributes defining a CEBM, offering a pathway for transitioning from traditional linear models to sustainable and regenerative practices. To achieve this, the study pursued four objectives: identifying factors shaping CEBM, exploring relevant frameworks and theories, determining influential attributes, and developing a comprehensive CEBM framework. A literature review and a Delphi study informed the findings. Key insights reveal that CEBM adoption is driven by attributes such as value propositions, client relationships, key resources, and take-back systems, all underpinned by established theories like systems thinking and institutional theory. Given its environmental impact and resource intensity, these elements are particularly relevant in the construction sector. The study's contribution extends to a validated CEBM framework, offering practical strategies for integration across procurement, design, construction, and deconstruction phases. By embracing circular principles, construction organisations can minimise waste, enhance resilience, and align with global sustainability goals, including the United Nations' Sustainable Development Goals. This study provides a vital roadmap for advancing sustainability in the built environment.

9.1 Introduction

This book aimed to establish a framework for a circular economy (CE) business model tailored for construction organisations. This framework identifies the key attributes defining a CE business model in this context. To achieve this overarching goal, the study pursued four specific objectives. A comprehensive literature review was conducted, followed by a Delphi study, to fulfil these objectives. The Delphi study validated findings from the literature review regarding the elements indicative of a CE business model. The subsequent section of this chapter presents the research's conclusion, which aligns with the study's objectives.

© The Author(s), under exclusive license to Springer Nature Switzerland AG 2025
O. Otasowie et al., *Circular Economy Business Model for Construction Organisations*, Green Energy and Technology, https://doi.org/10.1007/978-3-031-88322-4_9

9.2 Conclusion

9.2.1 Research Objective RO1

The study's first objective was to identify factors determining a CEBM for construction organisations. A comprehensive literature review was conducted to address this objective. The findings reveal that the CEBM for construction organisations is a complex framework shaped by multiple attributes rather than a single defining factor. Further exploration shows that the CEBM has gained increasing attention across various sectors, including manufacturing and the food industry. Moreover, it has been demonstrated that understanding the benefits of adopting a CEBM within the construction sector can significantly transform industry practices. These advancements can support construction organisations in establishing more effective operations, enhancing service delivery, and driving sustainability. Research also highlights that the CEBM aims to develop a regenerative and sustainable system by minimising waste, optimising resource utilisation, and ensuring resilience and longevity throughout the product life cycle. It also provides valuable insights for planning interventions and shaping policies. The literature underscores the CEBM as vital for organisations, offering an innovative alternative to the conventional linear "take, make, dispose" approach.

It is vital to note that while researchers have shown interest in the extent of CE model adoption across various industries, the construction sector's stakeholders and experts are positioned to play a crucial role in driving this shift. Key elements contributing to the CEBM include, but are not limited to, value proposition, client segment, channels, client relationships, key resources, key activities, key partnerships, cost structures, revenue streams, and take-back systems. The findings from the literature emphasise that addressing the challenge of creating more circular value propositions requires continued research and development efforts.

9.2.2 Research Objective RO2

The study's second objective was to identify frameworks and theories for CEBM. The exploration of frameworks and theories for CEBM reveals their pivotal role in driving the transition towards sustainable economic systems. The CEBM is not merely a structural innovation but a transformative approach to reimagining economic activities across industries, including the construction sector. CEBM challenges traditional linear models and drives long-term sustainability and resilience by emphasising regeneration, resource optimisation, and waste minimisation. The literature underscores that the development of CEBM frameworks is inherently multifaceted, integrating various dimensions such as value propositions, client relationships, key resources, and revenue streams. Established theories, including systems thinking, neoclassical theory, institutional theory, neo-Marxist theories, and the natural-resource-based view of the firm approach underpin these dimensions. Systems thinking, in particular, offers a holistic perspective, enabling organisations to understand interdependence and create regenerative solutions that align with environmental, social, and economic objectives. These theoretical underpinnings are foundational pillars for designing and implementing effective CEBM strategies.

The application of CEBM frameworks is especially significant in the construction industry due to the sector's substantial environmental impact and resource intensity. Adopting circular principles such as modular design, material reuse, and take-back systems can redefine industry practices, promoting sustainable construction methods. Moreover, CEBM frameworks support aligning organisational strategies with broader policy objectives, such as the United Nations' Sustainable Development Goals (SDGs) and climate action targets. In conclusion, frameworks and theories for CEBM provide a critical roadmap for achieving sustainable economic transformation. They empower organisations to create value through innovative business models while addressing global challenges such as resource scarcity and environmental degradation.

9.2.3 Research Objective RO3

The book's third objective was to determine the main and sub-attributes influencing CEBM for construction organisations. This study underscores the complexity and multifaceted nature of CEBM for construction organisations, emphasising that their implementation requires a holistic understanding of interconnected attributes. The findings demonstrate that the shift from traditional linear models to a CEBM framework is not driven by a single characteristic but rather by integrating multiple main and sub-attributes. The main attributes influencing CEBM in construction organisations include the value proposition, client segment, channels, client relationships, key resources, key activities, key partnerships, cost structure, revenue streams, take-back systems, material information management system, and organisational culture. These attributes form the foundation for developing sustainable and regenerative business practices, enabling construction organisations to minimise waste, optimise resource utilisation, and enhance resilience throughout the product lifecycle. Sub-attributes within these main categories further illuminate specific factors that drive the adoption of CEBM. In conclusion, implementing CEBM by construction organisations is a promising pathway toward achieving sustainability and resilience in the built environment.

9.2.4 Research Objective RO4

The book's fourth objective was to develop a CEBM framework for construction organisations. The development of a framework for a CEBM for construction organisations represents a significant step towards driving sustainability and resource efficiency within the construction sector. This study achieved this objective by identifying and validating key attributes that define a CEBM while aligning these findings with the unique characteristics and challenges of the construction industry. Through a detailed literature review, the study identified the essential elements of a CEBM, including value proposition, client segment, channels, client relationship, key resources, key activities, key partnerships, cost structure, revenue streams, take-back systems, material information management system, and organisational culture. These elements were critically analysed to determine their relevance and applicability to construction organisations, which face unique complexities due to their dependence on finite resources, large-scale operations, and extensive stakeholder networks.

A Delphi study was conducted to refine further and validate the literature review findings. This method allowed for expert consensus on the determinant attributes of a CEBM for construction organisations. The Delphi process reinforced the literature review findings and provided insights into practical implementation strategies, highlighting the importance of integrating circular practices into procurement, design, construction, and deconstruction phases. The study's contribution extends beyond theoretical insights by offering a structured framework that construction organisations can adopt to transition from traditional linear to circular models. This framework emphasises strategic enablers such as business innovation, collaboration across the supply chain, and policy alignment to facilitate the adoption of CE principles.

9.3 Summary

The book's conclusion is presented in this chapter. It presented the conclusion based on the study's objectives.

Index

© The Editor(s) (if applicable) and The Author(s), under exclusive license to Springer Nature Switzerland AG 2025

O. Otasowie et al., *Circular Economy Business Model for Construction Organisations*, Green Energy and Technology, https://doi.org/10.1007/978-3-031-88322-4

K

L

M

N

O

P

Q

R

The manufacturer's authorised representative in the EU is Springer Nature Customer Service Centre GmbH, Europaplatz 3, 69115 Heidelberg, Germany. If you have any concerns regarding our products, please contact ProductSafety@springernature.com

Printed and bound by CPI Group (UK) Ltd, Croydon, CR0 4YY

01/10/2026

02225934-0004